Topics on Strong Gravity

*A Modern View on
Theories and Experiments*

Other Related Titles from World Scientific

Centennial of General Relativity: A Celebration
edited by César Augusto Zen Vasconcellos
ISBN: 978-981-4699-65-5

Modified Gravity: Progresses and Outlook of Theories, Numerical Techniques and Observational Tests
edited by Baojiu Li and Kazuya Koyama
ISBN: 978-981-3273-99-3

Advanced Interferometric Gravitational-Wave Detectors (In 2 Volumes)
Volume I: Essentials of Gravitational-Wave Detectors
Volume II: Advanced LIGO, Advanced Virgo and Beyond
edited by David Reitze, Peter Saulson and Hartmut Grote
ISBN: 978-981-3146-07-5 (Set)
ISBN: 978-981-3146-10-5 (Vol. I)
ISBN: 978-981-3146-11-2 (Vol. II)

Loop Quantum Gravity for Everyone
by Rodolfo Gambini and Jorge Pullin
ISBN: 978-981-121-195-9

Topics on Strong Gravity

A Modern View on Theories and Experiments

Editor

César Augusto Zen Vasconcellos

Universidade Federal do Rio Grande do Sul, Brazil
International Center for Relativistic Astrophysics Network
Coordinating Center (ICRANet), Italy

World Scientific

NEW JERSEY · LONDON · SINGAPORE · BEIJING · SHANGHAI · HONG KONG · TAIPEI · CHENNAI · TOKYO

Published by

World Scientific Publishing Co. Pte. Ltd.

5 Toh Tuck Link, Singapore 596224

USA office: 27 Warren Street, Suite 401-402, Hackensack, NJ 07601

UK office: 57 Shelton Street, Covent Garden, London WC2H 9HE

Library of Congress Cataloging-in-Publication Data
Names: Vasconcellos, César A. Z., editor.
Title: Topics on strong gravity : a modern view on theories and experiments /
    editor, César Augusto Zen Vasconcellos.
Description: New Jersey : World Scientific, [2020] | Includes bibliographical references.
Identifiers: LCCN 2019030499 | ISBN 9789813277335 (hardcover)
Subjects: LCSH: Gravitation. | General relativity (Physics) | Astrophysics.
Classification: LCC QC178 .T67 2020 | DDC 521/.1--dc23
LC record available at https://lccn.loc.gov/2019030499

British Library Cataloguing-in-Publication Data
A catalogue record for this book is available from the British Library.

For any available supplementary material, please visit
https://www.worldscientific.com/worldscibooks/10.1142/11186#t=suppl

Desk Editor: Ng Kah Fee

Typeset by Stallion Press
Email: enquiries@stallionpress.com

To my wife Mônica, to my daughter Helena, to my son Marcio, to my stepdaughters Daniela, Marina and Barbara, and to Fiorella, with love.

Preface

Albert Einstein's theory of General Relativity (GR) is the most successful and profound theoretical description of the gravitational interaction. Despite the success of the theory, as is well known, there have been discussions, especially in more recent years, about the completeness of General Relativity. Numerous alternative and extended formulations or proposed modifications of General Relativity have been presented, as well as new equipments, laboratories and observatories have been built or planned in order to base the incessant search for answers to doubts and questions concerning this incompleteness. The regime of observation of these proposals, so that they can be properly tested, involves strong gravitational fields and high energy domains, that is, the regime in which space-time is highly curved, as in the surroundings of neutron stars, pulsars and black holes, and where a completely satisfactory quantum description of the gravitational field still remains elusive. These aspects have motivated the organization of this book that discusses some of the most important and current topics about strong gravity and GR related theories.

César Augusto Zen Vasconcellos
Porto Alegre, Brazil
November 2019

Introduction

For more than a century, our main understanding about gravity has been based on Albert Einstein's theory of General Relativity (GR). Einstein's theory fundamentally changed our understanding about the universe, about its origin and evolution. General Relativity accurately describes many phenomena occurring on very distinct scales. However, most of the GR-compatible observational and experimental results fall within the gravitational weak-field regime. In recent years, discrepancies between the data and the corresponding General Relativity predictions have been observed and have generated intense research activity. One of the most critical aspects of General Relativity refers to the prediction of spacetime singularities in extreme physical situations, that is, in the regime of strong gravity, where spacetime is highly curved. These discrepancies indicate that either the parameters of General Relativity should be reviewed in the regime of strong-field gravity and large spacetime curvature, or the theory itself should be modified.

In this book we concentrate our attention to extended alternative theories of gravity and to the best astrophysical laboratories to probe the strong gravity-field regime: black holes, pulsars and neutron stars.

About the Content of the Book

In the following we briefly describe the content of the chapters of the book.

Chapter 1. In this chapter, a few formal aspects about the main motivation of the book are presented. As stressed by C. Will,[1] "General Relativity works very well for gravity of ordinary strength, the variety experienced by humans on Earth or by planets as they orbit the sun. But it's never been tested in extremely strong fields, regions that lie at the boundaries of physics." In this regime, General Relativity may exhibit a spacetime singularity, i.e., a point in the *fabric of spacetime* in which all physical laws are indistinguishable from one another and where space and time are no longer interrelated realities, but merge indistinguishably and cease to have any independent meaning. In this chapter, we review the derivation of Friedmann equations and briefly discuss the presence of a singularity at time $t = 0$ (assumed to correspond to the beginning of the universe) since the components of the Riemann–Christoffel curvature tensor and the Ricci scalar curvature tend to infinity.

Chapter 2. In writing out this chapter, Daniela Pérez and Gustavo E. Romero were motivated by limitations of General Relativity (GR) and related models in the strong field regime, as for instance, the emergence of singularities in various spacetime models of astrophysical and cosmological importance, as well as the apparent incompatibility between GR and the unitary time evolution of quantum systems in the study of black hole thermodynamics, among other problems. In this chapter, the authors outline predictions of two families of theories that deviate from General Relativity in different aspects: $f(R)$-gravity and scalar–tensor–vector gravity (STVG). They discuss astrophysical effects in models based upon both matter and vacuum solutions. In particular, they analyse characteristics of the structure of neutron stars and the constraints on the parameters of the theories introduced by the latest observations in this field of research. They also review black hole solutions and several astrophysical consequences of them, including accretion disks and jets. Finally, they report on the implications of the detection of various gravitational wave events for these theories.

---

[1]Will, C.M., *Living Rev. Relativity* **17**, 4 (2014).

Chapter 3. The contribution of Peter O. Hess and Thomas Boller contains a review on the *pseudo-complex General Relativity* (pcGR) and its observational consequences. According to the authors, significant differences of the pcGR predictions in comparison to the corresponding General Relativity (GR) predictions appear only in the regime of strong gravitational fields. As an important aspect of their analysis, robust differences between the predictions of GR and pcGR are related to the description of structures formed by diffused material in orbital motion around a massive central body (accretion disks). The authors bring to this book an interesting discussion about extremely relevant stellar objects in the study of the effects of strong gravity on matter, i.e., black holes. Finally, some words are addressed by the authors on the observation of gravitational waves and about the corresponding predictions raised by pseudo-complex General Relativity.

Chapter 4. In this chapter, William M. Spinella and Fridolin Weber focus their attention to other relevant stellar objects in the study of the effects of strong gravity on matter: neutron stars. In the first part of their contribution they present a review on the relativistic mean-field approximation for determining the equation of state of hadronic matter along with a number of associated nuclear parameterizations. Then, constraints on the neutron star equation of state from nuclear physics and neutron star observations are discussed and used to eliminate several popular nuclear parameterizations. Finally, the authors determine the hadronic equation of state with the inclusion of hyperons and systematically investigate the parameter space range of the meson–hyperon coupling constants.

Chapter 5. In this chapter, Tomaso M. Belloni discusses important aspects about black holes binaries. Accordingly, in the past decades, the phenomenology of fast time variations of high-energy flux from black hole binaries has increased, thanks to the availability of more sophisticated space observatories, and a complex picture has emerged. Recently, according to him, models have been developed to interpret the observed signals in terms of fundamental frequencies connected to General Relativity (GR), and have opened this way a promising path to measure prediction of GR in the strong-field regime. Motivated by these theoretical and observational advances, Belloni presents in his contribution, a review about the current standpoint of black hole binaries from both the observational and theoretical points of view. As a result of his analysis, black hole binaries emerge as the most promising laboratories for testing GR and the observations available today suggest that the next observational facilities can lead to a great development in this area of research.

Chapter 6. In this chapter, José Antonio de Freitas Pacheco approaches the topic of supermassive black holes, that is, the regime of extreme super-strong gravity. According to him, the discovery of high redshift quasars represents a challenge in understanding the origin of supermassive black holes. In his contribution, two evolutionary scenarios are considered. The first one concerns massive black holes in the local universe, which in a large majority have been formed by the growth of seeds as their host galaxies are assembled in accordance with the hierarchical picture. In the second scenario, seeds with masses around 100–$150\,M_\odot$ grow by accretion of gas forming a non-steady massive disk, whose existence is supported by the detection of huge amounts of gas and dust in high-z quasars. These models of non-steady self-gravitating disks explain quite well the observed luminosity–mass relation of quasars at high-z, indicating also that these objects do not radiate at the so-called Eddington limit.

Chapter 7. In this chapter, David Blaschke, David Edwin Alvarez-Castillo, Alexander Ayriyan, Hovik Grigorian, Noshad Khosravi Largani, and Fridolin Weber introduce an effective equation of state (EoS) model based on polytropes, which serves to study the so-called "mass twins" scenario, where two compact stars have approximately the same mass but (significant for observation) quite different radii. Stellar mass twin configurations are obtained if a strong first-order phase transition occurs in the interior of a compact star. In the mass–radius diagram of compact stars, this will lead to a third branch of gravitationally stable stars with features that are very distinctive from those of white dwarfs and neutron stars. This scenario opens a new door in the study of compact stars.

We hope these chapters can contribute to broadening our readers' vision on the central theme of this book, strong gravity. And we also hope our readers share the same satisfaction we experienced as we organized this book. Good reading.

The Editor

Contents

5. Probing the Spacetime Around a Black Hole
with X-Ray Variability 153

Tomaso M. Belloni

List of Figures

List of Tables

Chapter 1

Spacetime Singularities in General Relativity

César A. Zen Vasconcellos

Instituto de Física
Universidade Federal do Rio Grande do Sul (UFRGS)
Av. Bento Gonçalves, 9500, Agronomia, Porto Alegre
RS, 91501-970, Brazil
cesarzen@cesarzen.com

Albert Einstein's theory of General Relativity provides the best description of gravitational interactions and has achieved over time countless successes in describing the most diverse events in the universe. However, as is well known, there has been a series of discussions in academic circles, especially in more recent years, about the incompleteness of General Relativity and predictions about the presence of singularities in the theory at extreme physical conditions. In this chapter, we review the derivation of Friedmann equations and briefly discuss the presence of a singularity at time $t = 0$ (assumed to correspond to the beginning of the universe) since the components of the Riemann–Christoffel curvature tensor and the Ricci scalar curvature tend to infinity.

1.1. General Relativity

The General Theory of Relativity (GR), developed by Albert Einstein (1916), one of the most important theoretical achievements in modern science, combines mechanics, gravitation and geometry in very elegant and concise covariant field equations, the Einstein equations, that describe gravity as a manifestation of the curvature of spacetime.

1

The theory of General Relativity is based on three main concepts: the Einstein equivalence principle (EEP), local Lorentz invariance (LLI) and local position invariance (LPI)[1] [Will (2014)]. The full symmetry of Special Relativity corresponds to the Poincaré symmetry group and includes the Abelian Lie group (P) of translations in space and time, the non-Abelian Lie group (J) of three-dimensional rotations in space and boosts, transformations connecting two uniformly moving bodies (K); J and K symmetries correspond to the Lorentz group.

As a consequence of the validity of the EEP, gravitation must correspond to a *metric theory*, in which (a) gravity is a manifestation of the curvature of a spacetime manifold endowed with a symmetric metric, (b) matter responds only to the geometry of spacetime, (c) the trajectories of freely falling test bodies correspond to geodesics of that metric and (d) for local freely falling reference frames, the non-gravitational laws of physics are covered in the framework of Special Relativity.[2]

Thus, in General Relativity, spacetime geometry determines free-falling world-lines due to the influence of matter, and from there

---

[1]In 1687, Sir Isaac Newton established the principle known today as the weak equivalence principle (WEP) which states that the mass of a body is proportional to its weight and he established this way the universality of free fall of massive bodies. The Einstein equivalence principle (EEP), on the other hand, is a more comprehensive concept, since it not only establishes the validity of the WEP, but also that the outcome of any local non-gravitational experiment is independent of the speed of the reference system in free fall (LLI) as well as from its spacetime location in the universe (LPI). EEP states this way the *strict equivalence between gravity and inertial acceleration*; this means that the laws of motion of free-falling frames in gravitational fields are completely equivalent to the corresponding laws of motion in uniformly accelerating inertial frames.

[2]According to Norman K. Glendenning, "perhaps the beauty of Einstein theory can be attributed to the essentially simple but amazing answer it provides to a fundamental question: what meaning is attached to the absolute equality of inertial and gravitational masses? If all bodies move in gravitational fields in precisely the same way, no matter what their constitution or binding forces, then this means that their motion has nothing to do with their nature, but rather with the *nature of spacetime*. And if spacetime determines the motion of bodies, then according to the notion of action and reaction, this implies that spacetime in turn is *shaped by bodies and their motion*." [Glendenning (2017)]

we may derive the metric of spacetime: matter moves according to the topological curvature of spacetime and matter in turn curves spacetime. For a given distribution of matter, the set of field equations of General Relativity allow the calculation of the spacetime metric, which in turn characterises the amount of curvature of spacetime. Accordingly, any test particle will move, under the action of gravity, along a geodesic of a symmetric metric defined by the metric tensor, $g_{\mu\nu}$, which in turn characterises the geometry of spacetime.

Since its formulation, about a hundred years ago, the continuous search for experimental confirmation of the General Relativity assumptions and for the establishment of its limits of validity, has been recurrent. The numerous efforts accordingly performed to test the predictions of the theory obtained especially from the 60s an extraordinary impulse with the discovery of quasars, pulsars, the cosmic background radiation, and black holes; the determination of the decrease in the orbital period of the Hulse–Taylor binary pulsar at a rate consistent with the General Relativity predictions of gravitational-wave energy loss; the detection of gravitational waves; among many others.

Historically, two distinct evolutionary stages have confronted the primary view of General Relativity, the quest for strong gravity — corresponding to a highly curved or dynamical spacetime[3] — and the quantization of gravity.

An analysis of the fundamental constants of the Special and General Theory of Relativity and quantum theory reveals a natural length scale at which quantum gravity effects can be expected to become important: the Planck length scale,

$$\ell_P = \sqrt{\frac{hG}{2\pi c^3}} = 1.6 \times 10^{-35}\,\text{m},$$

which defines a point of confluence between the natural constants that characterize gravity, G, and the quantum effects, h and c.

---

[3]In quantitative terms, the boundaries between strong and weak gravity can be established, albeit roughly speaking, by the quantity $\varepsilon \sim GM/Rc^s$; in this context, weak gravity would correspond to $\varepsilon \ll 1$.

Similarly we can define a time scale,

$$t_P = \frac{\ell_P}{c} = \sqrt{\frac{hG}{2\pi c^5}} = 5.4 \times 10^{-44}\,\text{s},$$

and a mass scale,

$$m_P = \sqrt{\frac{ch}{2\pi G}} = 22\,\mu\text{g}.$$

These space, time and mass scales are the results solely of the dimensional combination of the values of the constants that appear naturally in our physical laws. In this context, those quantities represent natural length, time and mass scales of quantum gravity; according to the principle of Special Relativity, these quantities would be Lorentz invariant, i.e., different observers should observe them to be the same. However, there are doubts on how to reconcile the supposed invariance of the physical quantities ℓ_P, t_P, m_P with Lorentz time dilation and length contraction. From the point of view of particle physics, there are aspects still unclear about the meaning of these scales as the Planck energy

$$E_P = \sqrt{\frac{hc^5}{2\pi G}}$$

equals 1.2×10^{31} eV, i.e., 10^{19} times higher than the maximum energy presently achieved by modern accelerators. Accordingly, the universe at the Planck time corresponds to a domain of strong gravity [Will (2014)].

GR allows in principle the existence of different kind of gravitational fields as for instance Brans–Dicke scalar fields, vector fields, background metric tensors and many others which interact with the metric tensor $g_{\mu\nu}$ directly. Moreover, accordingly to the principles of General Relativity, since the metric solely determines the motion of test bodies, these fields cannot couple directly to matter fields.

GR, although not the only relativistic theory which describes gravity, is the simplest theory consistent with experimental and observational data.[4] However, there are a number of issues still open with respect to GR, which remain over time. The most fundamental aspect in this regard is the understanding of how GR can be reconciled with the laws of quantum physics to generate a complete and self-consistent theory of quantum gravity.

According to the cosmological principle, our universe can be considered for very large scales as homogeneous and isotropic. In this domain, the Friedmann–Lemaître–Robertson–Walker metric (FLRW) [Friedmann (1922); Lemaître (1927); Robertson (1935); Walker (1937)] represents an exact solution of Einstein field equations of GR, describing this way a homogeneous, isotropic universe. According to the FLRW model, the universe must be, at any time, either in an expansion or contraction state.

In this regard, Hubble's observations have shown for the first time that the recessional velocity of galaxies increases with their distance from the Earth, implying the universe is expanding [Hubble (1926, 1929)]. Moreover, in 1999, observations showed that the universe is expanding at an increasing rate (accelerated expansion of the universe) [Perlmutter *et al.* (1999)]. Accordingly, the universe can go through a process of expansion, cooling and diffusion from a hot,

---

[4]Among all existing gravitational theories, the empirical evidence supporting the Einstein equivalence principle leads to the conclusion that the only viable theories of gravity are metric theories, or possibly theories which are metric apart from very weak or short-range non-metric couplings (as in string theory). Nowadays, we speak of the strong equivalence principle (SEP), which differs from EEP due to the inclusion of bodies with self-gravitational interactions (planets, stars) and of experiments involving gravitational forces (Cavendish experiments, gravimeter measurements). Notice that SEP contains the EEP as a special case in which the local gravitational forces are ignored [Will (2014)]. Additionally, nowadays we speak of two fundamental classes of metric theories, "purely dynamical" and "prior-geometric", i.e., metric theories for which gravitational fields have their structure and evolution determined by coupled partial differential field equations and metric theories that contain "absolute elements", fields or equations whose structure and evolution are given *a priori*, and are independent of the structure and evolution of the other fields of the theory. GR belongs to the second class of metric theories [Will (2014)].

dense and rapidly expanding past, at 13.7 billion years ago, to a cool, diffuse and slowly expanding present. GR suggests this way an universe in which as we go *backwards* in time, the hotter it was, the denser, and the more rapidly it was expanding and that at 13.7 billion years ago, at the extreme gravitational regimen, the density, temperature and expansion rate of the universe start off as infinite.

This represents a drastic consequence of General Relativity, the possibility that space and time may exhibit spacetime singularities, i.e., a point in which all physical laws are indistinguishable from one another, where space and time are no longer interrelated realities, but merge indistinguishably and cease to have any independent meaning. The prediction of singularities is undoubtedly a limitation of GR in the sense that one cannot expect that its classical description shall remain valid at the extreme physical conditions near a spacetime singularity.

These extreme conditions of the initial state of the universe are very far from our experimental possibilities and present theoretical models allow therefore only speculations about the avoidance of physical singularities or about the physical conditions that circumvented this drastic consequence of General Relativity. The main speculations in this regard explore the possibilities that the primordial spacetime singularity of the universe may have been avoided through quantization effects of gravitation or by a combination of effects associated with some sort of unification of General Relativity and quantum theory.

Speculation aside, in this book we focus our attention on the discussion of limitations of General Relativity in the limit of strong gravity, on alternative models and on implications of these limitations and alternative models in the description of compact stars, more precisely, neutron stars and black holes.

1.2. Einstein Equations of General Relativity

Schematically, Einstein equations are second rank tensor covariant field equations representing the topological geometric structure and the mass–energy content of spacetime with two kinds of fields, matter fields Φ and the metric/gravitational field $g_{\mu\nu}$.

In this context, a major challenge in GR has been to develop coherent technical treatments of Einstein equations for practical problems using tensor calculus, differential geometry and the theory of partial differential equations, without losing sight at the same time of the geometrical aspects of these equations. As a consequence, numerous physical predictions have resulted from the formal treatment of Einstein equations, as for example, the existence of tidal forces, gravitational waves, gravitational collapse, and the Big Bang cosmology

Einstein equations can be derived from the variational principle

$$\delta \mathcal{A}_{EH} = \delta \int \mathcal{L}_{EH} \, d^4 x = \delta \int \sqrt{-g} \, R \, d^4 x = 0, \qquad (1.1)$$

and the generally covariant formalism provided by Riemann geometry (equivalence of arbitrarily moving frames).

In expression (1.1) the Einstein–Hilbert action, $\mathcal{A}_{EH}$, is defined in terms of the invariant Einstein–Hilbert Lagrangian density, $\mathcal{L}_{EH}$. In this expression $\sqrt{-g}$ represents the scalar density,[5] with $g = \det(g_{\mu\nu})$, and where $g_{\mu\nu}$ denotes the metric tensor, which characterises the geometry of spacetime. In expression (1.1) R denotes the Ricci

---

[5]The four-volume element transforms in GR, under a coordinate change, as

$$dx'^0 dx'^1 dx'^2 dx'^3 = J dx^0 dx^1 dx^2 dx^3,$$

where J is the Jacobian of the transformation, $J = \det \left| \frac{\partial x'^\rho}{\partial x^\mu} \right|$. The corresponding transformation law for the metric tensor is

$$g_{\mu\nu} = \frac{\partial x'^\alpha}{\partial x^\mu} g'_{\alpha\beta} \frac{\partial x'^\beta}{\partial x^\nu},$$

and the determinant equation is $g = Jg'J = J^2 g'$ where $g = \det |g_{\mu\nu}|$ is a negative quantity (as can be seen from the Minkowski metric), leading to the expression $\sqrt{-g} = J\sqrt{-g'}$. Since $R = R'$ is a scalar field, then

$$\int_{V_4} R\sqrt{-g} \, d^4 x = \int_{V_4} R\sqrt{-g'} \, J d^4 x = \int_{V_4} R'\sqrt{-g'} \, d^4 x',$$

is an invariant quantity. The integral of the scalar density $R \equiv R\sqrt{-g}$ over a region of spacetime is invariant with respect to a spacetime coordinate transformation and $\sqrt{-g} \, d^4 x$ is the invariant volume element.

scalar, the only independent scalar constructed from the spacetime metric[6] as

$$R = g^{\mu\nu} R_{\mu\nu},   \tag{1.2}$$

where $R_{\mu\nu}$ defines the Ricci curvature tensor. The metric tensor $g_{\mu\nu}$ contains 10 different components and is the sole dynamical field or variable in GR; it describes gravity as the geometry of the spacetime while the Ricci tensor and the Ricci scalar measure its curvature. GR contains no arbitrary functions or parameters, except of course Newton's universal constant of gravitation G. The more general form of Einstein field equations is [Einstein (1916)]

$$G_{\mu\nu} + \Lambda g_{\mu\nu} = \frac{8\pi G}{c^4} T_{\mu\nu},   \tag{1.3}$$

with

$$G_{\mu\nu} = R_{\mu\nu} - \frac{1}{2} g_{\mu\nu} R,   \tag{1.4}$$

where Λ is the cosmological constant[7] and $T_{\mu\nu}$ is the divergence-less energy–momentum tensor (stress–energy tensor) which describes the properties of the spacetime geometry and encompasses all physical quantities (except gravity) that contribute to the energy content of spacetime. More precisely it describes the distribution of energy, momentum, and stresses associated with matter, radiation, and all sorts of force fields. Due to its symmetric properties, the

---

[6]Functions which depend on the first derivative or higher derivatives of the metric are not considered because they vanish for every point P which lies on an n-dimensional manifold $\mathcal{M}$ locally Euclidean, R^n. The property of being locally Euclidean is preserved by local homeomorphism.

[7]Recent observation of supernova indicates an accelerating universe, which would imply a non-zero cosmological constant or a dynamical dark energy contributing about 70 percent of the critical density. The cosmological constant has significance for quantum field theory, quantum gravity, and cosmology. On the scale of the solar system or of stellar systems its effects are negligible, for the values of the cosmological constant inferred from supernova observations [Will (2014)].

Einstein tensor $G_{\mu\nu}$ satisfies the four differential Bianchi identities, $(G_{\mu\nu})^{;\nu} = 0$ (one identity for each coordinate) in accordance with the (covariant) energy–momentum conservation law. Hence, only six of the 10 components of $G_{\mu\nu}$ are independent; thus the metric tensor, which is not an observable, is not completely determined from the stress–energy tensor; additionally, to determine it one needs four *coordinate conditions* which correspond in GR roughly speaking to the gauge fixing condition in electrodynamics.

1.3. Singularities at the Beginning of the Universe

The pioneering work of Vesto Slipher [Slipher (1913, 1917)], by measuring Doppler shifts of distant galaxies, was the starting point for the discovery by Edwin Powell Hubble and Milton Lasalle Humason [Hubble (1926, 1929)], of the expansion of the universe. These achievements allowed Hubble to conclude that the universe was expanding according to Hubble's law, a statement of a direct correlation between the distance to a galaxy and its recessional velocity as determined by redshift measurements.

They thus confirmed the conjectures presented a few years earlier by Georges Lemaître [Lemaître (1927)], who proposed an expanding model for the universe — to explain the observed redshifts of spiral nebulae — and who has predicted prior to Hubble, the law that would later be known as Hubble's law. Moreover, based on the work of Albert Einstein [Einstein (1916)] and Willem de Sitter [de Sitter (1916)], independently, Georges Lemaître derived the equations that would later be known as Friedmann's equations for an expanding universe. A few years later Georges Lemaître created the hypothesis of the primeval atom, today known as the Big Band theory.

1.4. Standard Cosmological Model

There has been a long debate about the relevance and role played by the scientists involved in the primordial steps of the development of GR, concerning observational research activities and the corresponding theoretical interpretations with regard to the comprehension of the dynamics of the universe.[8]

In the model of General Relativity developed in 1917, Albert Einstein introduced a cosmological constant to achieve a static (background) distribution of matter. In 1927, a non-static solution to the field equations, including matter sources, was proposed by Georges Lemaître, related to the already available observations of recession velocities of distant sources, which is interpreted nowadays as an evidence for the expansion of the universe. A few years before, in 1922, Alexander Friedmann already presented solutions to the General Relativity field equations with a time-dependent scale factor, including matter, but his solutions were not fully acknowledged until Lemaître's proposition.

Based on the cosmological principle, which modernly summarizes the notion that the distribution of matter in the universe is homogeneous and isotropic when observed on a large scale,[9] Alexander Friedmann [Friedmann (1922)], Georges Lemaître [Lemaître (1927)], Howard Robertson [Robertson (1935)] and Arthur Walker [Walker (1937)] developed what later became the standard model of cosmology, which is characterized by a maximally symmetric metric, in which a scale factor $a(t)$ accounts for the expansion or the collapse of the universe

$$ds^2 = dt^2 - a(t)^2 \left[\frac{dr^2}{(1 - kr^2)} + r^2(d\theta^2 + \sin^2\theta d\phi^2) \right], \qquad (1.5)$$

---

[8]There is a plethora of scientists who are systematically cited in this debate as, for example, Alexander Friedmann, Howard Robertson, Arthur Walker, Vesto Slipher, Milton La Salle Humason, Henrietta Swan Leavitt, Knut Lundmark, Georges Lemaître and Edwin Powell Hubble, among others [Trimble (1996)]. We should also add the names of: Hermann Minkowski, Carl Friedrich Gauss, Bernhard Riemann, Elwin Bruno Christoffel, Gregorio Ricci-Curbastro, Tulio Levi-Civita, Marcel Grossmann, David Hilbert, William Kingdon Clifford, Gunnar Nordström, and Paul Gerber.

[9]This is because it is expected that the forces of nature act uniformly throughout the universe, and should therefore not produce observable irregularities in the large-scale structure of the universe during its evolutionary process.

with the parameter k encoding the spatial curvature of the universe, $k = -1, 0, 1$ for, respectively, negatively, flat or positively curved spatial hyper-surfaces. Expression (1.5) corresponds to the FLRW metric which describes an isotropic and homogeneous universe, i.e., there is no any privileged direction and the symmetry of the universe is spherical.

1.5. Friedmann Equations

The Friedmann equations may be derived from the Einstein equations. Since the FLRW metric is diagonal, for the temporal part of Einstein equations we have:

$$R_{00} - \frac{1}{2}Rg_{00} + \Lambda g_{00} = 8\pi GT_{00}. \tag{1.6}$$

Similarly, for the spatial part we obtain:

$$R_{ii} - \frac{1}{2}Rg_{ii} + \Lambda g_{ii} = 8\pi GT_{ii}. \tag{1.7}$$

To determine the Friedmann equations we need the corresponding expressions for the Christoffel symbols of the FLRW metric, the Riemann–Christoffel curvature tensor and the energy–momentum tensor of a perfect fluid.

The equation of motion of a falling particle in an arbitrary frame considering an arbitrary gravitational field is

$$\frac{d^2x^\lambda}{d\tau^2} + \Gamma^\lambda_{\mu\nu}\frac{dx^\mu}{d\tau}\frac{dx^\nu}{d\tau} = 0. \tag{1.8}$$

Here $\Gamma^\lambda_{\mu\nu}$, defined by

$$\Gamma^\lambda_{\mu\nu} \equiv \frac{\partial x^\lambda}{\partial \xi^\alpha}\frac{\partial^2 \xi^\alpha}{\partial x_\mu \partial x_\nu}, \tag{1.9}$$

represents the *affine connection*, which is symmetric in its lower indices. In this expression x^μ and x^ν represent arbitrary coordinates in the gravitational field and ξ^α represents a coordinate in the free fall referential. According to this equation, the particle follows a geodesic spacetime path. Moreover, in the new and arbitrary reference frame, the second term of (1.8) causes a deviation from a straight-line motion of the particle. Therefore, the second term represents the effect of the gravitational field. Expressing Eq. (1.9) in another coordinate system, x'^μ, and using the chain rule several times to rewrite it, the affine connection $\Gamma^\lambda_{\mu\nu}$ may be defined as

$$\Gamma'^\lambda_{\mu\nu} = \frac{\partial x'^\lambda}{\partial x^\rho}\frac{\partial x^\tau}{\partial x'_\mu}\frac{\partial x^\sigma}{\partial x'_\nu}\Gamma^\rho_{\tau\sigma} + \frac{\partial x'^\lambda}{\partial x^\rho}\frac{\partial^2 x^\rho}{\partial x'_\mu x'_\nu}. \qquad (1.10)$$

According to the transformation laws of tensors, the presence of the second term on the right hand side of this expression causes the affine connection to be a non-tensor. The affine connection can be expressed in terms of the metric tensor and its derivatives which makes the spacetime metric behave as the gravitational potential and the related affine connection as the gravitational force. The procedure to achieve that goal is to use the fact that the difference between the affine connection and the Christoffel symbol of the second kind

$$\left\{ \begin{matrix} \lambda \\ \mu\nu \end{matrix} \right\} = \frac{1}{2}g^{\lambda\kappa}\left[\frac{\partial g_{\kappa\nu}}{\partial x^\mu} + \frac{\partial g_{\kappa\mu}}{\partial x^\nu} - \frac{\partial g_{\mu\nu}}{\partial x^\kappa}\right], \qquad (1.11)$$

is a tensor which vanishes — according to the equivalence principle — wherever and whenever the effects of gravitation, in a local inertial frame ξ^α, are absent. The Christoffel symbol provides this way a mean of computing the affine connection from the derivatives of the metric tensor, as well as to compute the Ricci tensor and the Ricci scalar appearing in expressions (1.6) and (1.7).

Taking into account that the FLRW metric is diagonal and has a symmetric connection, the Christoffel symbols which are different from zero are:

$$\Gamma^t_{rr} = \frac{a(t)\dot{a}(t)}{1 - kr^2};$$

$$\Gamma^t_{\phi\phi} = a(t)\dot{a}(t)r^2;$$

$$\Gamma^t_{\theta\theta} = a(t)\dot{a}(t)r^2\sin^2\theta;$$

$$\Gamma^r_{tr} = \Gamma^r_{rt} = \Gamma^\theta_{t\theta} = \Gamma^\theta_{\theta t} = \Gamma^\phi_{t\phi} = \Gamma^\phi_{\phi t} = \frac{\dot{a}(t)}{a(t)};$$

$$\Gamma^r_{rr} = \frac{r}{k^{-1}(1 - kr^2)};$$

$$\Gamma^r_{\theta\theta} = -r(1 - kr^2);$$

$$\Gamma^r_{\phi\phi} = -r(1 - kr^2)\sin^2\theta;$$

$$\Gamma^\theta_{r\theta} = \Gamma^\theta_{\theta r} = \Gamma^\phi_{r\phi} = \Gamma^\phi_{\phi r} = \frac{1}{r};$$

$$\Gamma^\theta_{\phi\phi} = -\sin\theta\cos\theta;$$

$$\Gamma^\phi_{\phi\theta} = \Gamma^\phi_{\theta\phi} = \frac{1}{\tan\theta}. \tag{1.12}$$

The Riemann–Christoffel curvature tensor — the only one in General Relativity that can be constructed from the metric tensor and its first and second derivatives —,

$$R^\rho_{\sigma\mu\nu} \equiv \Gamma^\rho_{\sigma\nu,\mu} - \Gamma^\rho_{\sigma\mu,\nu} + \Gamma^\alpha_{\sigma\nu}\Gamma^\rho_{\alpha\mu} - \Gamma^\alpha_{\sigma\mu}\Gamma^\rho_{\alpha\nu}, \tag{1.13}$$

allows to express the Ricci tensor, by contracting over two of the indices, in terms of the Christoffel symbols:

$$R_{\mu\nu} = \Gamma^\alpha_{\mu\alpha,\nu} - \Gamma^\alpha_{\mu\nu,\alpha} - \Gamma^\alpha_{\mu\nu}\Gamma^\beta_{\alpha\beta} + \Gamma^\alpha_{\mu\beta}\Gamma^\beta_{\nu\alpha}. \tag{1.14}$$

Using the previous results, the components of the Ricci tensor which are different from zero are:

$$R_{tt} = R^\kappa_{t\kappa t} = R^r_{trt} + R^\theta_{t\theta t} + R^\phi_{t\phi t} = 3\frac{\ddot{a}(t)}{a(t)};$$

$$R_{rr} = R^\tau_{r\tau r} = -\frac{1}{1 - kr^2}(a(t)\ddot{a}(t) + 2\dot{a}(t)^2 + 2k);$$

$$R_{\theta\theta} = R^\gamma_{\theta\gamma\theta} = -r^2 a(t)\ddot{a}(t) + 2r^2\dot{a}^2 + 2kr^2;$$

$$R_{\phi\phi} = R^\iota_{\phi\iota\phi} = -r^2 a(t)\ddot{a}(t)\sin^2\theta + 2r^2\dot{a}(t)^2\sin^2\theta$$
$$+ 2kr^2\sin^2\theta, \tag{1.15}$$

with $\kappa = r, \theta, \phi$, $\tau = t, \theta, \phi$, $\gamma = r, t, \phi$, and $\iota = r, t, \theta$. Summarizing the results, we get a diagonal Ricci tensor with the time (R_{tt}) and space (R_{rr}) components

$$R_{tt} = 3\frac{\ddot{a}(t)}{a(t)};$$

$$R_{ii} = R^{\tau}_{r\tau r} + R^{\gamma}_{\theta\gamma\theta} + R^{\iota}_{r\iota r}$$

$$= \frac{g_{ii}}{a(t)^2}(a(t)\ddot{a}(t) + 2\dot{a}(t)^2 + 2k). \tag{1.16}$$

For the Ricci scalar curvature we obtain:

$$R = g^{\mu\nu}R_{\mu\nu} = 6\left[\left(\frac{\ddot{a}(t)}{a(t)}\right) + \left(\frac{\dot{a}(t)}{a(t)}\right)^2 + \frac{k}{a(t)^2}\right]. \tag{1.17}$$

A perfect fluid corresponds to a medium in which the pressure is isotropic in the rest frame of each fluid element, and shear stresses and heat transport are absent. An observer with a velocity $\mathbf{v}$ at a certain point of the fluid will observe the fluid in the neighborhood as isotropic with an energy density ε and pressure p. In this local frame the energy–momentum tensor is

$$T'^{\mu\nu} \equiv \begin{pmatrix} \varepsilon & 0 & 0 & 0 \\ 0 & p & 0 & 0 \\ 0 & 0 & p & 0 \\ 0 & 0 & 0 & p \end{pmatrix}. \tag{1.18}$$

As viewed from an arbitrary frame, for instance the laboratory system, let this fluid element be observed to have velocity $\mathbf{v}$. According to the inverse Lorentz transformation

$$x^{\mu} = \Lambda_{\nu}{}^{\mu}x'^{\nu}, \tag{1.19}$$

where $\Lambda_{\nu}{}^{\mu}$ characterizes the matrix transformation, we obtain the tensor transformation

$$T^{\mu\nu} = \Lambda_{\alpha}{}^{\mu}\Lambda_{\beta}{}^{\nu}T'^{\alpha\beta}. \tag{1.20}$$

The elements of the inverse transformation are given as

$$\Lambda_0{}^0 = \Lambda_1{}^1 = \gamma,$$

$$\Lambda_1{}^0 = \Lambda_0{}^1 = v\gamma, \tag{1.21}$$

$$\Lambda_2{}^2 = \Lambda_3{}^3 = 1,$$

or, by taking a boost in an arbitrary direction with the primed axis having velocity $\mathbf{v} = (\mathbf{v}^1, \mathbf{v}^2, \mathbf{v}^3)$ relative to the unprimed, as

$$\Lambda_0{}^0 = \gamma,$$
$$\Lambda_0{}^j = \Lambda_j{}^0 = v^j\gamma, \qquad (1.22)$$
$$\Lambda_j{}^k = \Lambda_k{}^j = \delta_k^j + (\gamma - 1)v^j v^k/\mathbf{v}^2.$$

In the arbitrary frame we get

$$T^{\mu\nu} = -p\eta^{\mu\nu} + (p + \varepsilon)u^\mu u^\nu; \qquad (1.23)$$

this expression reduces to the diagonal form above when $\mathbf{v} = 0$.

From these equations the corresponding expressions for the T_{00} and T_{ii} components of the energy–momentum tensor, to be inserted in expressions (1.6) and (1.7), read[10]:

$$T_{00} = \rho g_{00}; \quad T_{ii} = -p g_{ii}, \qquad (1.24)$$

with $g_{\mu\nu}$ replaced by $\eta_{\mu\nu}$ which represents the Minkowski (flat) spacetime metric tensor, given in rectilinear coordinates by

$$\eta_{\mu\nu} \equiv \begin{pmatrix} 1 & 0 & 0 & 0 \\ 0 & -1 & 0 & 0 \\ 0 & 0 & -1 & 0 \\ 0 & 0 & 0 & -1 \end{pmatrix}, \qquad (1.25)$$

since isotropy and homogeneity of spacetime leads to a diagonal metric tensor. From these equations we obtain, for a perfect fluid

$$T = T_\alpha^\alpha = \rho - 3p; \qquad (1.26)$$

in this expression, we follow Einstein summation convention over the set of index represented by the symbol α.

---

[10]Since a perfect fluid is isotropic, the macroscopic speed of the fluid cannot have a privileged direction, so it has only a temporal component: $u^\alpha = (1, 0, 0, 0)$.

Combining these results with the General Relativity equations, for a homogeneous and isotropic universe there are only two independent Friedmann equations:

$$\left(\frac{\dot{a}(t)}{a(t)}\right)^2 = \frac{8\pi G}{3}\rho(t) + \frac{\Lambda}{3} - \frac{k}{a^2(t)}, \tag{1.27}$$

and

$$\left(\frac{\ddot{a}(t)}{a(t)}\right) = -\frac{4\pi G}{3}(\rho(t) + 3p(t)) + \frac{\Lambda}{3}. \tag{1.28}$$

The Friedmann equations state that in general conditions the universe is not static.

Combining these expressions, for the particular case $k = 0$, Friedmann equations reduce to

$$3\left(\frac{\rho(t) + p(t)}{a(t)}\right) + \frac{\dot{\rho}(t)}{\dot{a}(t)} = 0. \tag{1.29}$$

Friedmann equations correspond to the set of equations that govern the expansion of spacetime in homogeneous and isotropic models of the universe within the context of General Relativity. Since $a(t)^2 \to 0$ as $t \to 0$, from Eqs. (1.16) and (1.17), there is a singularity at $t = 0$ (assumed to correspond to the beginning of the universe) because the components of the Riemann–Christoffel curvature tensor and the Ricci scalar curvature tend to infinity:

$$R = g^{\mu\nu}R_{\mu\nu} \longrightarrow \infty, \tag{1.30}$$

and consequently these quantities are not continuously defined at $t = 0$. The predictions of Friedmann equations lose all physical meaning in this regime where all physical laws are indistinguishable from one another and where space and time are no longer interrelated realities, but merge indistinguishably and cease to have any independent meaning. This essential singularity at $t = 0$ cannot be transformed away by any coordinate transformation. In this sense it is considered a real singularity. The existence of real singularities where the curvature scalars and densities diverge implies that all physical laws break down. In general, if the determinant of $g_{\mu\nu}$ tends to zero, the usual Riemannian invariants diverge, and the covariant derivative is not well defined, because the inverse of the metric becomes singular. This makes the Christoffel symbols of the second kind, as well as the Riemann–Christoffel curvature tensor, the Einstein tensor, the Ricci tensor and the scalar curvature to become singular.

The possible existence of singularities as predicted by Einstein equations and the consequent limitations of their predictions in the regime of strong gravity was the main motivation to organise this book. Numerous theoretical, experimental and observational efforts are being carried out today to offer alternative models that can answer the many questions raised by the possible incompleteness of General Relativity. This book gives the readers a modern insight into some of the theoretical and experimental advances being made to answer these questions.

References

Churchill, R. and Brown, J., in *Complex Variables and Applications* (McGraw-Hill, New York, 1990).

de Sitter, W., *Mon. Not. Roy. Astron. Soc.* **76**, 699 (1916).

Einstein, A., *Annalen der Physik* **49**, 769 (1916); **51**, 639 (1916). Einstein, A., *Preussische Akademie der Wissenschaften, Sitzungsberichte* **Part 1**, 423 (1916); **Part 1**, 688 (1916); **Part 2**, 1111 (1916).

Friedmann, F., *Zeitschrift für Physik* **10**, 377 (1922).

Glendenning, N.K., in *Centennial of General Relativity: A Celebration*, Ed. C.A. Zen Vasconcellos (World Scientific, Singapore, 2017).

Hubble, E., *Astrophys. J.* **64**, 321 (1926).

Hubble, E., *Proc. Nat. Acad. Sci.* **15**, 168 (1929).

Lemaître, G., *Annales de la Société Scientifique de Bruxelles* **47**, 49 (1927).

Linde, A., *Lect. Notes Phys.* **738**, 1 (2008).

Mantz, C.L.M. and Prokopec, T., *Foundations of Physics* **41**, 1597 (2011).

Perlmutter, S. *et al.*, *Astrophys. J.* **517** (2), 565 (1999).

Robertson, H.P., *Astrophys. J.* **82**, 284 (1935).

Slipher, V.M., *Lowell Observatory Bulletin* **2**, 56 (1913).

Slipher, V.M., *Proc. Am. Philos. Soc.* **56**, 403 (1917).

Trimble, V., *Pub. Astron. Soc. Pacific* **108** (730), 1073 (1996).

Walker, A.G., *Proceedings of the London Mathematical Society* **2** (1), 90 (1937).

Will, C.M., *Living Rev. Relativity* **17**, 4 (2014).

Chapter 2

Astrophysical Constraints on Strong Modified Gravity

Daniela Pérez* and Gustavo E. Romero[†,‡]

Instituto Argentino de Radioastronomía
Centro Científico Tecnológico (CCT)
Consejo Nacional de Investigaciones
Científicas y Técnicas (CONICET)
Comisión de Investigaciones Científicas de la
Provincia de Buenos Aires (CICPBA)
C.C.5, 1894 Villa Elisa
Buenos Aires, Argentina
**danielaperez@iar-conicet.gov.ar*
[†]*romero@iar-conicet.gov.ar*

We offer a discussion on the strong field regime predictions of two families of theories that deviate from General Relativity in different aspects: $f(R)$-gravity and scalar–tensor–vector gravity (STVG). We discuss astrophysical effects in models based upon both matter and vacuum solutions of such theories. In particular, we analyze neutron star structure and the constraints on the parameters of the theories introduced by the latest observations. We also review black hole solutions and several of their astrophysical consequences, including accretion disks and jets. Finally, we report on the implications of the detection of various gravitational wave events for these theories.

[‡]Also at Facultad de Ciencias Astronómicas y Geofísicas, Universidad Nacional de La Plata (UNLP), Paseo del Bosque s/n, 1900 La Plata, Buenos Aires, Argentina.

2.1. General Relativity in the Strong Field Domain: Problems and Challenges

General Relativity (GR) is a theory of space, time, and gravitation formulated by Albert Einstein [Einstein (1916)]. In the theory spacetime is considered as an entity endowed with physical properties. Its physical geometry is represented by a continuous and differentiable 4-dimensional pseudo-Riemannian manifold M with a metric field $g_{\mu\nu}$. The geometric properties of the manifold are related to the different matter fields existing in spacetime by Einstein's field equations:

$$R_{\mu\nu} - \frac{1}{2}Rg_{\mu\nu} = \frac{8\pi G}{c^4}T_{\mu\nu}, \qquad (2.1)$$

where $R_{\mu\nu}$ is the Ricci tensor obtained from the contraction of the Riemann curvature tensor, R is the Ricci (or curvature) scalar, and $T_{\mu\nu}$ is the energy–momentum of all material fields. This is a set of ten nonlinear hyperbolic-elliptic partial differential equations in the coefficients of the metric field. Solving the equations for some distribution of energy and momentum, one can determine the free motion of test particles through the geodetic equation:

$$\frac{d^2x^\lambda}{ds^2} + \Gamma^\lambda_{\mu\nu}\frac{dx^\mu}{ds}\frac{dx^\nu}{ds} = 0, \qquad (2.2)$$

where $ds^2 = g_{\mu\nu}dx^\mu dx^\nu$ is the spacetime interval and $\Gamma^\lambda_{\mu\nu}$ is the *affine connection* of the manifold:

$$\Gamma^\lambda_{\mu\nu} = \frac{1}{2}g^{\lambda\alpha}(\partial_\mu g_{\nu\alpha} + \partial_\nu g_{\mu\alpha} - \partial_\alpha g_{\mu\nu}). \qquad (2.3)$$

In GR, then, the effects of gravitation are a consequence of the curvature of spacetime. These effects exist in a non-local way, since curvature always vanishes on sufficiently small scales.

GR works wonderfully in the weak field regime. It passes all tests performed in the solar system [Will (2014)] and provides an adequate description of most astrophysical phenomena. The theory correctly predicts the value for the perihelion advance of Mercury and the bending of light around the Sun. Gravitational redshift, another classic prediction of the theory, has been successfully tested with different experiments. Shapiro delay was also confirmed with high confidence (a parameter $g = 1.000021 \pm 0.000023$, against a value of $g = 1$ for GR). Frame dragging and the geodetic effect have also been confirmed. The strong equivalence principle has been tested to $h = 4.4 \times 10^{-4}$, with $h = 0$ in GR. Gravitational lensing has also confirmed GR to better than 1%. More radically, the prediction of the existence of gravitational waves was spectacularly verified in 2015 with the detection of the first black hole binary merger event, dubbed GW150914 [Abbott *et al.* (2016a)].

Despite all these successes, GR is not free of problems. Singularities naturally appear in the theory for a number of spacetime models of both astrophysical [Penrose (1965)] and cosmological importance [Hawking and Penrose (1970)]. In addition, black hole thermodynamics seems to suggest an incompatibility between GR and unitary evolution of quantum systems [Giddings (1995)]. When applied to large scales, as those of galaxies and clusters of galaxies, GR requires the inclusion of mysterious dark matter to explain rotation curves and galaxy velocities. And when applied to the universe as a whole, a strange field of "dark energy" must be assumed to account for the observed accelerated expansion. Also, spacetime is expected to have quantum properties at the Planck scale, but GR is not renormalizable and therefore cannot be used to make meaningful physical predictions on such scales.

Not surprisingly, many attempts have been made at producing new spacetime theories that might overcome some of these problems. These theories are generally called "modified gravity". They should be almost identical to GR on the scales of the solar system, where Einstein's theory is in accord with the experiments with exquisite accuracy. They might differ, however, in some not well-explored domains, such as large scales and in the strong gravity regime. In this article we shall explore some of the constraints imposed by current astrophysical observations upon some of these theories.

2.2. Modified Gravity: Different Approaches

If a particle moves in spacetime departing from the expected geodetic trajectory, we face one out of three possibilities: 1) The trajectory actually is not a geodetic one and there are non-gravitational
fields acting on the particle; 2) The trajectory is geodetic but there
is matter not taken into account in our energy–momentum tensor;
or 3) The trajectory is geodetic but the law that relates spacetime properties with energy–momentum is not correctly described
by Eq. (2.1). Situation 1 applies, for instance, to a charged particle affected by a magnetic field. Situation 2 requires a modification of the ontology accepted by our theory. This is the case when
we keep Einstein's field equations untouched but we introduce dark
matter or dark fields to explain rotation galactic curves, gravitational lensing, or the accelerated expansion of the universe. Option
3 is one of ontological parsimony: in order to keep our ontology at a
minimum we adopt a new prescription for the way spacetime inter-
acts with other fields. This latter path demands modifications into
the left side of Einstein's field equations, i.e. in the properties of
spacetime. Such changes, of course, should be subtle enough as to
yield the same predictions obtained from the original equations in
those domains where the theory has passed stringent tests. The new
solutions, however, might differ substantially from those of GR in
the little known strong regime of high curvature or on cosmological
scales.

There are a number of ways to modify Eq. (2.1) in order to achieve
such effects. Einstein himself championed these attempts since 1917
till his death. His first modification consisted in introducing a cosmological term linear in the metric. This changes Eq. (2.1) into:

$$R_{\mu\nu} - \frac{1}{2}Rg_{\mu\nu} + \Lambda g_{\mu\nu} = \frac{8\pi G}{c^4}T_{\mu\nu}. \qquad (2.4)$$

Here, Λ is the so-called cosmological constant. The effect of the new
term is to allow for repulsive gravity over some scales. Einstein fine
tuned Λ to obtain a static (latter shown to be unstable) solution.
Currently, such a term with a constant of value $\Lambda = 1.11 \times 10^{-52}$
m^{-2} is used in the standard cosmological model that includes cold

dark matter (CDM), the ΛCDM model. In such a model the accelerated expansion is accounted for by the gravitational repulsion experienced by the cosmic fluid over some critical size. The same result can be obtained replacing the cosmological term on the left side of the equations by a positive vacuum energy density on the right side. Such energy density is called "dark energy". Notice that these are two different approaches that yield the same result: in the first case we modify the law of gravitation; in the second one, we add a dark field with a peculiar equation of state of the form $p = -\rho$.

A general way of introducing changes in the geometric sector of Eq. (2.1) is to change the relativistic action. GR is obtained from the action:

$$S[g] = \int \frac{1}{2\kappa} R \sqrt{-g} \, d^4x. \tag{2.5}$$

This action can be generalized to

$$S[g] = \int \frac{1}{2\kappa} f(R) \sqrt{-g} \, d^4x, \tag{2.6}$$

where g is the determinant of the metric tensor and $f(R)$ is some function of the curvature (Ricci) scalar.

The generalized field equations are obtained by varying with respect to the metric. The variation of the determinant is

$$\delta \left(\sqrt{-g} \right) = -\frac{1}{2} \sqrt{-g} \, g_{\mu\nu} \delta g^{\mu\nu}. \tag{2.7}$$

The Ricci scalar is defined as

$$R = g^{\mu\nu} R_{\mu\nu}. \tag{2.8}$$

Therefore, its variation with respect to the inverse metric $g^{\mu\nu}$ is given by

$$\delta R = R_{\mu\nu} \delta g^{\mu\nu} + g^{\mu\nu} \delta R_{\mu\nu}$$
$$= R_{\mu\nu} \delta g^{\mu\nu} + g^{\mu\nu} (\nabla_\rho \delta \Gamma^\rho_{\nu\mu} - \nabla_\nu \delta \Gamma^\rho_{\rho\mu}). \tag{2.9}$$

Since $\delta \Gamma^\lambda_{\mu\nu}$ is actually the difference of two connections, it should transform as a tensor. Therefore, it can be written as

$$\delta \Gamma^\lambda_{\mu\nu} = \frac{1}{2} g^{\lambda a} (\nabla_\mu \delta g_{a\nu} + \nabla_\nu \delta g_{a\mu} - \nabla_a \delta g_{\mu\nu}), \tag{2.10}$$

and substituting in the equation above we get:

$$\delta R = R_{\mu\nu}\delta g^{\mu\nu} + g_{\mu\nu}\Box\delta g^{\mu\nu} - \nabla_\mu\nabla_\nu\delta g^{\mu\nu}. \tag{2.11}$$

The variation in the action results in:

$$\delta S[g] = \frac{1}{2\kappa}\int \left(\delta f(R)\sqrt{-g} + f(R)\,\delta\sqrt{-g}\right)\mathrm{d}^4x$$

$$= \frac{1}{2\kappa}\int \left(F(R)\,\delta R\sqrt{-g} - \frac{1}{2}\sqrt{-g}\,g_{\mu\nu}\delta g^{\mu\nu}\,f(R)\right)\mathrm{d}^4x$$

$$= \frac{1}{2\kappa}\int \sqrt{-g}\left[F(R)(R_{\mu\nu}\delta g^{\mu\nu} + g_{\mu\nu}\Box\delta g^{\mu\nu} - \nabla_\mu\nabla_\nu\delta g^{\mu\nu})\right.$$

$$\left. - \frac{1}{2}g_{\mu\nu}\,\delta g^{\mu\nu}\,f(R)\right]\mathrm{d}^4x, \tag{2.12}$$

where $F(R) = \frac{\partial f(R)}{\partial R}$. Integrating by parts on the second and third terms we get

$$\delta S[g] = \frac{1}{2\kappa}\int \sqrt{-g}\,\delta g^{\mu\nu}\left[F(R)R_{\mu\nu} - \frac{1}{2}g_{\mu\nu}f(R)\right.$$

$$\left. + (g_{\mu\nu}\Box - \nabla_\mu\nabla_\nu)F(R)\right]\mathrm{d}^4x. \tag{2.13}$$

By demanding that the action remains invariant under variations of the metric, i.e. $\delta S[g] = 0$, we find the field equations in generic $f(R)$-gravity:

$$F(R)R_{\mu\nu} - \frac{1}{2}f(R)g_{\mu\nu} + [g_{\mu\nu}\Box - \nabla_\mu\nabla_\nu]F(R) = \kappa T_{\mu\nu}, \tag{2.14}$$

where $T_{\mu\nu}$ is the energy–momentum tensor defined as

$$T_{\mu\nu} = -\frac{2}{\sqrt{-g}}\frac{\delta(\sqrt{-g}\,L_{\mathrm{m}})}{\delta g^{\mu\nu}},$$

and L_{m} is the matter Lagrangian. If $F(R) = 1$, i.e. $f(R) = R$, we recover Einstein's theory.

Equation (2.14) is a system of non-linear partial differential equations of four order in the coefficients of the metric tensor field $g_{\mu\nu}$. A full description of $f(R)$-gravity can be found in the book by Capozziello and Faraoni [Capozziello and Faraoni (2011)].

An interesting feature is that the Ricci scalar R and the trace of the energy–momentum tensor $T = T^\mu_\nu$ are related in a differential way. This implies that for some prescriptions of $f(R)$, the Ricci scalar can be different from zero even if $T = 0$.

In the case of constant curvature $R = R_0$, Eq. (2.14) in the absence of matter fields becomes:

$$R_{\mu\nu} = -\Lambda g_{\mu\nu},\qquad(2.15)$$

where

$$\Lambda = \frac{f(R_0)}{f'(R_0) - 1}.\qquad(2.16)$$

We have, then, that cosmological solutions with accelerated expansion can be obtained in $f(R)$-gravity without adopting a cosmological constant term or dark fields.

In GR, the effects of gravity are understood as the result of the curvature of spacetime. Such a curvature is described by the Riemann tensor, which consists of second order derivatives of the tensor metric field. A different approach to modified gravity consists in ascribing to spacetime not only tensor fields, but also scalar and vector aspects. One of such attempts is known as scalar–tensor–vector gravity (STVG) and has been presented by J. Moffat [Moffat (2006)].

In STVG theory, gravity is not only an interaction mediated by a tensor field, but by scalar and vector fields. The action of the full gravitational field is:

$$S = S_{\text{GR}} + S_\phi + S_{\text{S}} + S_{\text{M}},\qquad(2.17)$$

where

$$S_{\text{GR}} = \frac{1}{16\pi G} \int d^4x \sqrt{-g}\, R,\qquad(2.18)$$

$$S_\phi = -\omega \int d^4x \sqrt{-g}\left(\frac{1}{4}B^{\mu\nu}B_{\mu\nu} - \frac{1}{2}\mu^2\phi^\mu\phi_\mu\right),\qquad(2.19)$$

$$S_{\text{S}} = \int d^4x \sqrt{-g}\left[\frac{1}{G^3}\left(\frac{1}{2}g^{\mu\nu}\nabla_\mu G\nabla_\nu G + V(G)\right)\right.$$
$$\left. + \frac{1}{G\mu^2}\left(\frac{1}{2}g^{\mu\nu}\nabla_\mu\mu\nabla_\nu\mu + V(\mu)\right)\right].\qquad(2.20)$$

Here, $g_{\mu\nu}$ denotes the spacetime metric, R is the Ricci scalar, ∇_μ the covariant derivative, ϕ^μ denotes a Proca-type massive vector field, μ is the mass and $B_{\mu\nu} = \partial_\mu\phi_\nu - \partial_\nu\phi_\mu$. $V(G)$ and $V(\mu)$ denote possible potentials for the scalar fields $G(x)$ and $\mu(x)$, respectively. We adopt units such that $c = 1$. The term S_M refers to possible matter fields.

Varying the action with respect to $g^{\mu\nu}$ and doing some simplifications, the field equations result in

$$G_{\mu\nu} = 8\pi G(T^\mathrm{M}_{\mu\nu} + T^\phi_{\mu\nu}), \tag{2.21}$$

where $G_{\mu\nu}$ denotes the Einstein tensor, and $T^\mathrm{M}_{\mu\nu}, T^\phi_{\mu\nu}$ are the matter and vector field energy–momentum tensors, respectively. The enhanced gravitational coupling is $G = G_\mathrm{N}(1+\alpha)$, where G_N denotes Newton's gravitational constant, and α a free parameter. Notice that STVG coincides with GR for $\alpha = 0$.

Variation of the simplified action with respect to ϕ_μ yields:

$$\nabla_\nu B^{\mu\nu} = -\frac{\sqrt{\alpha G_\mathrm{N}}}{\omega} J^\mu, \tag{2.22}$$

where J^μ denotes the four-current matter density, and the constant $\sqrt{\alpha G_\mathrm{N}}$ is determined to adjust the phenomenology. This vector field is completely absent in GR.

Certainly $f(R)$ and STVG are not the only families of modified gravity available. There are literally hundreds of alternative theories of gravitation, including multidimensional theories, torsion theories, bi-metric theories, theories of variable speed of light, and many, many more. An exhaustive discussion of the strong regime of all of them largely exceeds what can be contained in a single book, not to mention a single chapter. Hence, we opt to choose two of them, that we deem represent two different and rather natural generalizations of GR. For more references on other approaches the reader can turn to the mentioned book by Capozziello and Faraoni [Capozziello and Faraoni (2011)].

How can we test the validity of theories such as STVG or $f(R)$-gravity? The answer is through their strong field effects. In this regime the fields behave differently from GR. Hence, studies of black holes, radiative effects in their surroundings, neutron stars, and other astrophysical objects where gravity is strong, are paramount to establish the validity of these theories well beyond the regime for which they were devised and originally applied. What remains of this chapter is devoted to discuss how the strong field regime astrophysics of compact objects changes if these theories were correct. Then, using the best available observational data we can impose some constraints of the validity domain of the theories.

2.3. Neutron Stars in Modified Gravity

2.3.1. *Introduction*

Neutron stars are among the most compact astrophysical objects in the universe. They are very dense stellar remnants where gravitational forces are balanced by neutron degeneracy pressure. Their masses are in the range of 1.5 $M_\odot$ to 2.2 $M_\odot$ and radii typically of the order of 10 kilometers, reaching a density in the inner core of approximately 8×10^{17} kg m^{-3}, i.e., 10^{14} times the density of iron. The intense gravitational field of neutron stars makes them ideal objects to test current theories of gravitation in the strong field domain.

The prediction of the existence of neutron stars was independent of astronomical observations. After the discovery of the neutron by Chadwick in 1932, Baade and Zwicky [Baade and Zwicky (1934)] were the first to suggest a new class of compact stars in which a core of degenerate neutrons could support the object against gravitational collapse.[1] In 1939, Oppenheimer and Volkoff [Oppenheimer and Volkoff (1939)] and Tolman [Tolman (1939)] produced the first neutron star model, assuming the star as an ideal neutron gas. They showed the existence of an upper limit for the stellar mass of $0.75 M_\odot$ above which the star is not longer stable and collapses into a black hole.

After the Second World War, the progress in the field of observational astronomy led to a series of discoveries that confirmed the theoretical predictions of previous decades. In 1967, a group

---

[1]In the same work, they developed a theory for supernova explosions and proposed that these explosions could be the origin of cosmic rays.

of astronomers headed by Anthony Hewish detected astronomical objects that emitted regular radio pulses [Hewish *et al.* (1968)]. In the same year, Shklovsky [Shklovsky (1967)] developed a detailed model to explain the radiation produced from Sco X-1, the first X-ray binary ever detected [Giacconi *et al.* (1962); Sandage *et al.* (1966)]. In his model, Shklovsky correctly established that the radiation was produced by the accretion of gas from a donor star onto a neutron star. The first binary pulsar was discovered by Hulse and Taylor in 1974.[2] Pulsars are rapidly spinning neutron stars whose strong magnetic field produces conical beams of electromagnetic radiation. If the axis of rotation of the neutron star does not coincide with the magnetic axis, external observers see the beams whenever the magnetic axis points towards them as the star rotates. The pulses have the same period as the neutron star.

Given their clock-like regularity and their compact nature, pulsars offer a natural laboratory for studying the gravitational field. The analysis of the signals from pulsars in binaries provides information on the properties of these systems. In particular, pulsar timing techniques have proven to be extremely useful for estimating relativistic deviations from Keplerian motions in the case of binary pulsars with high velocities and strong gravitational fields. The measurement of the five post-Keplerian (PK) parameters[3] made possible very precise tests of GR and alternative theories of gravity. See for instance Zhang and Saha (2017); Zhang *et al.* (2019); De Laurentis *et al.* (2017); Seymour and Yagi (2018).

The recent detection of gravitational wave (GW) signals from a stellar-mass binary system by the LIGO/Virgo detectors [Abbott *et al.* (2017c)], and the subsequent observation of a short γ-ray burst

---

[2]The pulsar mass was measured very precisely and it was found to be 1.44 $M_\odot$. The hypothesis of an ideal gas of neutrons for the interior of the star was ruled out, showing that the interactions between the nucleons needed to be considered [Camenzind (2007)].

[3]The five post-Keplerian (PK) parameters are: the rate of the periastron advance $\dot{\omega}$, the orbital period decay $\dot{P}_b$, the so-called relativistic γ (the Einstein term corresponding to time dilatation and gravitational redshift), and the Shapiro delay term r (range) and s (shape).

associated with the event made possible new approaches for studying the properties of neutron stars. Using the data from GW170817 and general relativistic magnetohydrodynamics simulations (GRMHD), upper limits to the maximum gravitational mass, $M_{\text{max}}^{\text{sph}}$, of a non-rotating, spherical neutron star were recently obtained [Ruiz, Shapiro, and Tsokaros (2018)].

Neutron stars (NSs) have been investigated in the framework of some alternative theories of gravitation: for instance in Horndenski gravity [Maselli *et al.* (2016)], beyond Horndenski theories [Babichev *et al.* (2016)], Einstein–Gauss–Bonnet–Dilaton theory [Pani *et al.* (2011); Kleihaus *et al.* (2016)], and Chern–Simons gravity [Yagi *et al.* (2013)]. In the following, we will focus on the main results for neutron star models in $f(R)$-gravity and STVG, emphasizing both the theoretical and observational predictions.

2.3.2. *Neutron star models in $f(R)$-gravity*

Compact objects have been largely studied in $f(R)$ theories. The first works focused on the derivation and solution of the Tolman–Oppenheimer–Volkoff (TOV) equations that describe a spherically symmetric mass distribution in hydrostatic equilibrium. In order to solve the field equations, a perturbative approach was adopted, considering the $f(R)$ function as a perturbation of a GR background (see, for instance, [Cooney, Dedeo, and Psaltis (2010); Arapoğlu, Deliduman, and Ekşi (2011); Deliduman, Ekşi, and Keleş (2012)]). Arapoğlu *et al.* [Arapoğlu, Deliduman, and Ekşi (2011)] and Deliduman *et al.* [Deliduman, Ekşi, and Keleş (2012)] used the perturbative approach and adopted a realistic equation of state (EoS) for the matter distribution in R-squared and $R_{\mu\nu}R^{\mu\nu}$ gravities, respectively.

The mass–radius relation obtained by these authors for such $f(R)$ models allowed for larger masses of NSs than those currently estimated for the most massive known pulsars: 2.01 ± 0.04 $M_\odot$ for J0348+0432 [Antoniadis *et al.* (2013)] and 1.928 ± 0.017 $M_\odot$ for J1614-2230 [Demorest *et al.* (2010)]. Orellana and coworkers [Orellana *et al.* (2013)] have also studied the R-squared case using a polytropic approximation for the EoS, and also a more realistic one.

The mass–radius relation obtained by these authors is consistent with the works previously mentioned: for the highest absolute values admitted for the free parameter of the theory, i.e., the α parameter, $f(R)$-gravity models predict higher masses of NSs than GR for every EoS.

An intriguing feature of the NS models developed by Orellana *et al.* is a mass profile that in some regions decreases with the radius. In GR this effect could only be explained by means of a fluid of negative density. This is not the case in $f(R)$-gravity, where the coupling between the spacetime geometry and the matter could naturally give rise to such effects. This particular result, however, should be considered with caution as the authors clearly stated, since it could be the consequence of the analytical representation of the EoS or the perturbative approach adopted.

Later, it was pointed out by Yazadjiev and coworkers [Yazadjiev *et al.* (2014)] that the use of the perturbative method to investigate the strong field regime in $f(R)$-theories may lead to unphysical results.[4] In order to obtain self-consistent models of NSs, they suggested to solve the field equations simultaneously, assuming appropriate boundary conditions. This approach was then applied by some authors [Astashenok, Capozziello, and Odintsov (2015); Capozziello *et al.* (2016); Aparicio Resco *et al.* (2016)].

The internal structure of NSs was also explored using the Palatini formalism [Kainulainen, Reijonen, and Sunhede (2007)]. In the Palatini formalism the metric and the connection are *a priori* considered independent geometrical objects.[5] The advantage of this formalism is that the field equations have derivatives of the metric up to second order, as in GR. Though the apparent mathematical simplification of

---

[4]Recently, Blázquez-Salcedo *et al.* [Blázquez-Salcedo *et al.* (2018)] investigated the axial quasi-normal modes of the neutron star model in R-squared gravity developed by Yazadjiev and coworkers [Yazadjiev *et al.* (2014)].

[5]In $f(R)$-gravity with a chameleon mechanism, Liu *et al.* [Liu, Zhang, and Zhao (2018)] found constraints for a general $f(R)$ function using observations of the pulsars PSR J0348+0432 and PSR J1738+0333. This restriction, however, is weaker than the one derived from the solar system observations.

the field equations, NS models in this formalism present serious short-comings as shown by Barause *et al.* in a series of two works [Barausse, Sotiriou, and Miller (2008a,b)]. For a polytropic equation of state, the authors demonstrated the appearance of divergences in the curvature invariants near the surface of the star, indicating that the origin of the singularity is related to the intrinsic features of Palatini $f(R)$-gravity [Barausse, Sotiriou, and Miller (2008b)]. For a realistic EoS, and choosing the $f(R)$ function as $f(R) = R + \alpha R^2$, Barausse and coworkers found that the radial profile of the mass parameter develops bumps when rapid changes in the derivatives of the EoS occur.

Motivated by the later result, Teppa Pannia *et al.* [Teppa Pannia *et al.* (2017)] aimed to investigate whether the non-smoothness of the mass parameter was rooted to the nature of $f(R)$-gravity in the Palatini formalism,[6] or was an effect of the particular EoS chosen.

The method used was to calculate the structure of a star in the Palatini formalism in R-squared gravity for two EoS: first an EoS similar to the one employed by Barausse *et al.* [Barausse, Sotiriou, and Miller (2008a)]; second, an EoS based on the connection of multiple polytropes that allows to control the derivatives of the EoS using a set of parameters. The results found were in accordance with those of Barausse *et al.* [Barausse, Sotiriou, and Miller (2008a,b)]: for both EoS, a) the maximum masses were lower than in GR, b) the mass profile displays regions where $dm/d\rho < 0$.

In conclusion, these investigations strongly suggest that the odd features in NS models in Palatini $f(R)$-gravity do not lay in the characteristics of the EoS nor in the particular mathematical method employed to solve the field equations, but are inherent to the nature of the theory.[7]

---

[6]Recently, Wojnar [Wojnar (2018)] analyzed the stability of neutron stars in Palatini $f(R)$-gravity.

[7]More complex models of NSs in $f(R)$-gravity have been developed which include the fluid anisotropy [Folomeev (2018)], rotation [Staykov *et al.* (2014); Yazadjiev *et al.* (2015)], magnetic fields [Cheoun *et al.* (2013); Astashenok, Capozziello, and Odintsov (2015); Bakirova and Folomeev (2016)], and different $f(R)$ functions and

2.3.3. *Neutron stars in scalar–tensor–vector gravity*

Contrary to $f(R)$-gravity, neutron star models in STVG have just started to be explored. Currently, there is only one work in the scientific literature on this issue by Lopez Armengol and Romero [Lopez Armengol and Romero (2017a)]. These authors derived the modified TOV equation assuming a static, spherically symmetric geometry for spacetime; the stellar matter content was modeled with a static, spherically symmetric, perfect fluid, energy momentum tensor.

Four distinct neutron EoS were considered: POLY [Silbar and Reddy (2004)], SLy [Douchin and Haensel (2001)], FPS [Pandhari-pande *et al.* (1989)] and BSK21 [Goriely, Chamel, and Pearson (2010); Pearson, Goriely, and Chamel (2011); Pearson *et al.* (2012)]. The first EoS is mathematically simple and well-behaved. The purpose of employing POLY to construct neutron star models was that if any particular feature arises in the model, it would probably be an effect of STVG. Conversely, SLy, FPS and BSK21 are realistic EoS.

When integrating numerically the equations, particular attention is needed for the free parameter α of the theory. The parameter plays a fundamental role since it mediates both the gravitational repulsion and enhanced gravitational attraction, being its value dependent on the mass source of the gravitational field. Moffat[8] determined for solar mass sources an upper limit given as [Moffat (2006)]:

$$\alpha_\odot \ll \frac{1.5 \times 10^5 \; c^2}{G_{\mathrm{N}}} \frac{1}{M_\odot} \; \mathrm{cm}. \qquad (2.23)$$

Here, G_{N} stands for the Newton gravitational constant, and c denotes the speed of light. Since neutron stars have few solar masses, the values of α were chosen to satisfy inequality (2.23). It was also taken into account that inside the star, α_{NS} would depend on the mass of each r-shell. A linear *ad hoc* prescription was defined to sample

---

formalisms [Alavirad and Weller (2013); Astashenok *et al.* (2017); Liu, Zhang, and Zhao (2018)].

[8]The restriction on α was imposed in order to find agreement of STVG predictions with the perihelion advance of Mercury.

different values of α:

$$\alpha_{NS} = \gamma \frac{1.5 \times 10^5 \; c^2}{G_N} \frac{1}{M_\odot} \left(\frac{M(r)}{M_\odot} \right) \quad \text{cm}, \qquad (2.24)$$

where $M(r)$ is the mass of the neutron star up to the r-shell, and γ is a normalized factor $\gamma \in [0; 1)$.

One of the main findings of Lopez Armengol and Romero [Lopez Armengol and Romero (2017a)] is that neutron star models in STVG admit higher total masses than in GR. This result deserves particular attention since recent determinations of neutron star masses defy GR limits [Antoniadis *et al.* (2013); Kiziltan *et al.* (2013); Özel *et al.* (2012); Demorest *et al.* (2010)]. The authors could also set a more restrictive upper limit for the parameter α by imposing within their model, neutron stars with realistic masses (in accordance with astronomical observations), and monotonically decreasing density profiles. The new restriction for the α parameter for stellar mass sources is:

$$\alpha < 10^{-2} \frac{1.5 \times 10^5 \; c^2}{G_N} \frac{1}{M_\odot} \quad \text{cm}. \qquad (2.25)$$

There are many issues that remain to be explored in relation to neutron stars in STVG; for instance, solutions that take into account the rotation of the star, the scalar field contributions, stability analysis, and quasi-normal modes of spherically symmetric solutions. STVG, thus, seems to be a rich field of research for us in the future.

2.4. Black Holes in Modified Gravity

Black holes constitute the most extreme manifestation of gravity. These objects are spacetime regions causally disconnected from the rest of the universe by an event horizon. Black holes were first postulated theoretically [Schwarzschild (1916)], and half a century later their astrophysical manifestation started to be detected [Hazard, Mackey, and Shimmins (1963); Webster and Murdin (1972); Bolton (1972)]. There is overwhelming astronomical evidence that supports their existence. The latest and most direct proof is the detection of gravitational waves produced by the merger of binary systems of black holes [Abbott *et al.* (2016a,c, 2017a,b,f)].

In the light of the recent discoveries, any viable alternative theory of gravitation should admit black hole solutions. This is the case for $f(R)$-gravity and STVG that now we will proceed to analyze.

2.4.1. *Black holes in $f(R)$-gravity*

There is an extensive literature on black hole solutions in $f(R)$-gravity; during the course of a decade, almost every year new solutions have been reported. This situation is in stark contrast with GR. Because of the uniqueness theorem, we know that the only stable stationary asymptotical black hole solutions in GR are within the Kerr–Newman family [Israel (1967, 1968); Carter (1971); Hawking (1972); Robinson (1975)]. In $f(R)$-gravity, however, the Birkoff theorem does not hold [Sotiriou and Faraoni (2010); Faraoni (2010)], and is still an open question whether "hairy" black hole solutions for a non-constant Ricci scalar exist in $f(R)$-gravity [Cañate, Jaime, and Salgado (2016); Cañate (2018); Sultana and Kazanas (2018)].

Black hole geometries have been found in $f(R)$-gravity both in the metric and Palatini formalisms.[9] As clearly stated by Sotiriou and Faraoni [Sotiriou and Faraoni (2010)], all black hole solutions of GR (with a cosmological constant) will also be solutions of $f(R)$ in both formalisms (see also [Psaltis (2008); Barausse and Sotiriou (2008)]). In the Palatini formalism, they will comprise the complete set of black hole solutions of the theory. In the metric formalism, the Birkoff theorem does not hold, thus, other black hole solutions can in principle exist.

In $f(R)$-gravity, the Ricci scalar can depend both on space and time. Given the extreme complexity of the field equations, a simplifying assumption is to consider the Ricci scalar constant $R = R_0$. Several black hole solutions were determined under this hypothesis: static spherically symmetric solutions [de La Cruz-Dombriz, Dobado,

---

[9]For black hole solutions in Palatini formalism see, for instance, [Olmo and Rubiera-Garcia (2011, 2012a,b); Bazeia (2014); Olmo, Rubiera-Garcia and Sanchez-Puente (2015a)].

and Maroto (2009, 2011)], charged solutions [Moon, Myung, and Son (2011a)] (see also [Hendi, Panah, and Mousavi (2012)]) and the corresponding generalization in Kerr–Newman spacetimes [Cembranos, de la Cruz-Dombriz, and Jimeno Romero (2011)]; static spherically symmetric solutions coupled to linear and non-linear electromagnetic fields [Habib Mazharimousavi, Halilsoy, and Tahamtan (2012a)], and also minimally coupled to a non-linear Yang–Mills field [Habib Mazharimousavi, Halilsoy, and Tahamtan (2012b)]; higher dimensional ($d \geq 4$) charged black holes have also been explored [Sheykhi (2012)].

Without assuming $R = R_0$, Perez Bergliaffa and Chifarelli de Oliveira studied the necessary conditions for an $f(R)$ theory to have static spherically symmetric black hole solutions [Bergliaffa and Nunes (2011)]. Following the "near horizon test" introduced by the previous authors, Mazharimousavi and coworkers [Habib Mazharimousavi, Kerachian, and Halilsoy (2013)] derived the necessary conditions for the existence of Reissner–Nordström black holes in $f(R)$-gravity.

A formalism for the generation of spherically symmetric metrics in d-dimensions both in vacuum and in the presence of matter sources was given by Amaribi *et al.* [Amirabi, Halilsoy, and Mazharimousavi (2016)]. Gao and Shen [Gao and Shen (2016)] also proposed a new method to find exact solutions for static, spherically symmetric spacetimes in this theory.[10]

In what follows, we will describe the main properties of the $f(R)$–Kerr black hole with constant Ricci scalar since such spacetime has been employed to constrain the parameters of some class of $f(R)$ theories (see Sec. 2.5).

The axisymmetric, stationary, and constant Ricci scalar geometry that describes a black hole with mass, electric charge, and angular

---

[10]Further black hole solutions were studied in $f(R)$ with conformal anomaly [Hendi and Momeni (2011)]; regular $f(R)$ black hole solutions have also been explored (see for instance [Rodrigues (2016)]). For spherically symmetric charged black holes in a class of $f(R)$ theories see work by Nashed *et al.* [Nashed (2018); Nashed and Capozziello (2019)]. Stability analysis for different class of $f(R)$ black hole solutions can be found in [Myung, Moon, and Son (2011); Moon, Myung, and Son (2011b); Myung (2011)].

momentum was found by Carter [Carter (1973)]. This geometry was used to study $f(R)$ black holes by Cembranos and coworkers [Cembranos, de la Cruz-Dombriz, and Jimeno Romero (2011)]. The line element takes the form[11] (we have set $Q = 0$):

$$ds^2 = \frac{\rho^2}{\Delta_r}dr^2 + \frac{\rho^2}{\Delta_\theta}d\theta^2 + \frac{\Delta_\theta \sin\theta^2}{\rho^2}\left[a\frac{c\,dt}{\Xi} - \left(r^2 + a^2\right)\frac{d\phi}{\Xi}\right]^2$$
$$- \frac{\Delta_r}{\rho^2}\left(\frac{c\,dt}{\Xi} - a\,\sin\theta^2\frac{d\phi}{\Xi}\right), \tag{2.26}$$

where,

$$\Delta_\theta = 1 + \frac{R_0}{12}a^2\cos\theta^2, \tag{2.27}$$

$$\Delta_r = \left(r^2 + a^2\right)\left(1 - \frac{R_0}{12}r^2\right) - \frac{2GMr}{c^2}, \tag{2.28}$$

$$\rho^2 = r^2 + a^2\cos\theta^2, \tag{2.29}$$

$$\Xi = 1 + \frac{R_0}{12}a^2. \tag{2.30}$$

Here M and a denote the mass and angular momentum of the black hole, respectively. If $R_0 \to 0$, Eq. (2.26) represents the spacetime metric in GR as expected.[12]

The relation between the Ricci scalar R_0 and the $f(R)$ function can be derived from the field equations of the theory in the metric formalism:

$$R_{\mu\nu}\left(1 + f'(R)\right) - \frac{1}{2}g_{\mu\nu}\left(R + f(R)\right)$$
$$+ \left(\nabla_\mu\nabla_\nu - g_{\mu\nu}\Box\right)f'(R) + \frac{16\pi G}{c^4}T_{\mu\nu} = 0, \tag{2.31}$$

---

[11]The line element for an $f(R)$–Schwarzschild black hole can be obtained by setting $a = 0$.

[12]Calzà, Rinaldi and Sebastiani [Calzà, Rinaldi, and Sebastiani (2018)] studied spherically symmetric solutions in vacuum for a special class of $f(R)$ functions that satisfy $f(R_0) = 0$ and $df(R_0)/dR = 0$. Some of the metrics obtained represent spherically symmetric black holes. Under a specific choice of the values of some parameters, the $f(R)$–Schwarzschild black hole solution presented here is recovered.

where $R_{\mu\nu}$ is the Ricci tensor, $\Box \equiv \nabla_\beta \nabla^\beta$, $f'(R) = df(R)/dR$, and $T_{\mu\nu}$ is the energy–momentum tensor. If we take the trace of the latter equation, we obtain:

$$R\left(1 + f'(R)\right) - 2\left(R + f(R)\right) - 3\Box f'(R) + \frac{16\pi G}{c^4} T = 0. \quad (2.32)$$

In the case of constant Ricci scalar R_0 without matter sources, from Eq. (2.32) we get the relation we were looking for:

$$R_0 = \frac{2 \, f(R_0)}{f'(R_0) - 1}. \quad (2.33)$$

It also should be noticed that if $R = R_0$ and $T_{\mu\nu} = 0$, Eq. (2.14) can be re-cast as:

$$R_{\mu\nu} = -\Lambda g_{\mu\nu}, \quad (2.34)$$

and,

$$\Lambda = \frac{f(R_0)}{f'(R_0) - 1}. \quad (2.35)$$

From relations (2.33) and (2.35) we see that $f(R)$ with constant Ricci scalar and no matter sources is formally equivalent to GR with a cosmological constant. This equivalence, however, is not physical as some authors have recently concluded [Cañate, Jaime, and Salgado (2016)].

The location of the event horizon as a function of the radial coordinate is obtained by setting $1/g_{\mathrm{rr}} = 0$:

$$\Delta_{\mathrm{r}} = \left(r^2 + a^2\right)\left(1 - \frac{R_0}{12} r^2\right) - \frac{2GMr}{c^2}. \quad (2.36)$$

For a nearly maximally rotating black hole $a = 0.99$, such as Cyg X-1 [Gou (2011)], the existence (or absence) of horizons depends on the value of the Ricci Scalar[13] R_0. If $R_0 \in (0, 0.6]$, there are 3 event horizons: the inner and outer horizons of the black hole and a

---

[13]In what follows, we express the values of Ricci scalar as a dimensionless quantity: $R_0 \equiv R_0 r_{\mathrm{g}}^2$, where $r_{\mathrm{g}} = GM/c^2$.

cosmological horizon; for $R_0 > 0.6$ there is a cosmological horizon that becomes smaller for larger values of R_0. If $R_0 \in (-0.13, 0)$ there are 2 event horizons. For $R_0 \leq -0.13$ naked singularities occur [Pérez, Romero, and Perez Bergliaffa (2013)]. Black hole solutions, thus, occur for $R_0 \in (-0.13, 0.6]$.

2.4.2. *Black holes in scalar–tensor–vector gravity*

The only known black hole solutions in STVG were found by Moffat [Moffat (2015)]. These represent static, spherically symmetric and also stationary, axially symmetric black holes, and were derived under the following assumptions:

- The mass m_ϕ of the vector field ϕ^μ was neglected because its effects manifest at kiloparsec scales from the source.[14]
- G is a constant that depends on the parameter α [Moffat (2006)]:

$$G = G_N (1 + \alpha), \tag{2.37}$$

where G_N denotes Newton's gravitational constant, and α is a free dimensionless parameter.

Given these hypothesis, and after solving the STVG field equations, the line element of the spacetime metric of a black hole of mass M and angular momentum $J = aM$ in STVG theory is [Moffat (2015)]:

$$ds^2 = -c^2 \left(\Delta - a^2 \sin^2 \theta \right) \frac{dt^2}{\rho^2} + \frac{\rho^2}{\Delta} dr^2 + \rho^2 d\theta^2$$

$$+ \frac{2ac \sin^2 \theta}{\rho^2} \left[(r^2 + a^2) - \Delta \right] dt d\phi$$

$$+ \left[(r^2 + a^2)^2 - \Delta a^2 \sin^2 \theta \right] \frac{\sin^2 \theta}{\rho^2} d\phi^2, \tag{2.38}$$

---

[14]The mass of the field ϕ^μ determined from galaxy rotation curves and galactic cluster dynamics [Moffat and Rahvar (2013, 2014); Moffat and Toth (2015)] is $m_\phi = 0.042$ kpc^{-1}.

$$\Delta = r^2 - \frac{2GMr}{c^2} + a^2 + \frac{\alpha G_N G M^2}{c^4}$$

$$= r^2 - \frac{2G_N(1+\alpha)Mr}{c^2} + a^2 + \frac{\alpha(1+\alpha)G_N^2 M^2}{c^4}, \qquad (2.39)$$

$$\rho^2 = r^2 + a^2 \cos^2 \theta. \qquad (2.40)$$

The metric above reduces to the Kerr metric in GR when $\alpha = 0$. By setting $a = 0$ in Eq. (2.38), we recover the metric of a Schwarzschild STVG black hole.

The radius of the inner r_- and outer r_+ event horizons are determined by the roots of $\Delta = 0$:

$$r_\pm = \frac{G_N(1+\alpha)M}{c^2} \left\{ 1 \pm \sqrt{1 - \frac{a^2 c^4}{G_N^2(1+\alpha)M^2} - \frac{\alpha}{(1+\alpha)}} \right\}. \qquad (2.41)$$

If we set $\alpha = 0$ in the latter expression, we obtain the formula for the inner and outer horizons of a Kerr black hole in GR. Inspection of Eq. (2.41) also reveals that for $\alpha > 0$ the outer horizon of a Kerr black hole in STVG is larger than the corresponding one in GR.

The radial coordinate of the ergosphere is determined by the roots of $g_{tt} = 0$:

$$r_E = \frac{G_N M (1+\alpha)}{c^2} \left[1 \pm \sqrt{1 - \frac{a^2 \cos^2 \theta \, c^4}{G_N^2 (1+\alpha)^2 M^2} - \frac{\alpha}{1+\alpha}} \right]. \qquad (2.42)$$

We see that the ergosphere grows in size as the parameter α increases.

Astrophysical black holes strongly interact with the surrounding media giving rise to a wide variety of high energy phenomena; by studying the particles and radiation produced in these sources, the properties of the black hole spacetime can be inferred, and thus the underlying theory of gravitation can be put to the test.

In the coming two sections we provide a brief account of the salient features of accretion disks and relativistic jets around black holes in $f(R)$-gravity and STVG.

2.5. Accretion Disks

The process of matter with angular momentum falling into a black hole may lead to the formation of an accretion disk. The matter rotating around the black hole loses angular momentum because of the friction between adjacent layers and spirals inwards; in the process the kinetic energy of the plasma increases and the disk heats up emitting thermal energy.

The characterization of the nature of the turbulence in the disk, and hence of the dissipation mechanisms constitutes the main problem for the formulation of a consistent theory of accretion disks. Some simplifying assumptions, however, can be made in order to construct realistic models of disks. Next, we shall consider accretion disks in steady state where the accretion rate is an external parameter, and the turbulence is characterized by a single parameter α which was first introduced by Shakura [Shakura (1972)] and Shakura and Sunyaev [Shakura and Sunyaev (1973)].

Novikov, Page, and Thorne [Novikov and Thorne (1973); Page and Thorne (1974)] generalized the latter model to include the strong gravitational fields' effects. They assumed the background spacetime geometry to be stationary, axially symmetric, asymptotically flat, reflection-symmetric with respect to the equatorial plane, and the self-gravity of the disk was considered negligible. The central plane of the disk is located at the equatorial plane of the spacetime geometry, and since the disk is assumed to be thin[15] the metric coefficients only depend on the radial coordinate r.

From the relativistic equations for the conservation of mass, energy, and angular momentum, Page and Thorne [Page and Thorne (1974)] obtained three fundamental equations for the time-averaged

---

[15] $\Delta z = 2h \ll r$, with z the height above the equatorial plane, and h is the thickness of the disk at radius r.

radial structure of the disk.[16] In particular, they provided an expression for the heat emitted by the accretion disk that reads:

$$Q(r) = \frac{\dot{M}}{4\,\pi\sqrt{-g}} \frac{\Omega_{,r}}{(E^\dagger - \Omega L^\dagger)^2} \int_{r_{\text{isco}}}^{r} \left(E^\dagger - \Omega L^\dagger\right) L^\dagger_{,r}\, dr. \qquad (2.43)$$

Here, $\dot{M}$ stands for the mass accretion rate, g is the metric determinant, and r_{isco} denotes the radius of the innermost stable circular orbit. The expressions for the specific energy $E^\dagger$, specific angular momentum $L^\dagger$, and angular velocity Ω of the particles that move on equatorial geodesic orbits around the black hole are:

$$E^\dagger(r) \equiv -u_t(r), \qquad (2.44)$$

$$L^\dagger(r) \equiv u_\phi(r), \qquad (2.45)$$

$$\Omega(r) \equiv \frac{u^\phi}{u^t}, \qquad (2.46)$$

where $u_t(r)$ and $u_\phi(r)$ are the t and ϕ components of the four-velocity of the particle, respectively. Formulas (2.44), (2.45), and (2.46) can also be written in terms of the metric coefficients (see Harko *et al.* [Harko, Kovács, and Lobo (2009)] or Pérez *et al.* [Pérez, Romero, and Perez Bergliaffa (2013)] for the corresponding expressions).

In order to compute the heat emitted by a thin accretion disk around a black hole, we first have to study the circular orbits of massive particles in such spacetime geometry. There are several papers devoted to the investigation of geodesics in both $f(R)$-gravity and STVG. In particular, the analysis of stable circular orbits of massive particles in $f(R)$–Schwarzschild and $f(R)$–Kerr black holes[17] was

---

[16]The average is taken over a time interval Δt during which it is assumed that the external geometry of the hole is modified negligibly, but the accretion of matter for any radius r is large compared with the typical mass enclosed in a ring of thickness r.

[17]Since $f(R)$ with constant Ricci scalar and no matter sources is formally equivalent to GR with a cosmological constant (see Sec. 2.4.1), the results of Pérez *et al.* are in accordance with the analysis of circular orbits in Schwarzschild–de Sitter, Schwarzschild–anti-de Sitter done by Stuchlík and Hledík [Stuchlík and Hledík

performed by Pérez and coworkers [Pérez, Romero, and Perez Bergliaffa (2013)]. The existence and location of stable circular orbits depend on the value of the Ricci scalar R_0. For both Schwarzschild and Kerr black holes in $f(R)$-gravity, if $R_0 > 0$ there is a minimum (r_{isco}) and maximum (r_{osco}) radius for stable circular orbits; for $r > r_{\text{osco}}$ there are no stable circular orbits in such spacetime. On the contrary, for $R_0 < 0$ stable circular orbits begin from a minimum radius and extend to the rest of the spacetime. In Table 2.1, we show the values of the Ricci scalar for which stable circular orbits are possible (second column), and the corresponding r_{isco} and r_{osco} (columns 3 and 4, respectively) for $f(R)$–Schwarzschild and $f(R)$–Kerr spacetimes $(a = 0.99)$. The formulas from which these results were obtained can be found in [Pérez, Romero, and Perez Bergliaffa (2013)] and references therein.

The existence and location of stable circular orbits for Schwarzschild and Kerr black holes in STVG were studied in [Pérez, Armengol, and Romero (2017)]. The range of values adopted for the parameter α were chosen in order to model accretion disks around stellar and supermassive black holes. For stellar mass sources, $\alpha < 0.1$ was taken in accordance to the work of Lopez Armengol and Romero [Lopez Armengol and Romero (2017a)]. In the case of

Table 2.1   Values of the innermost and outermost stable circular orbits for $f(R)$–Schwarzschild and $f(R)$–Kerr spacetimes $(a = 0.99)$. The second column indicates the values of Ricci scalar R_0 for which stable circular orbits are possible. The values of r_{isco} and r_{osco} are expressed in gravitational radii.

Spacetime	R_0	r_{isco}	r_{osco}
$f(R)$ Schwarzschild	$R_0 \in (0, 2.85 \times 10^{-3})$ $R_0 < 0$	$r_{\text{isco}} \in (6, 7.5)$ $r_{\text{isco}} \in (3.75, +\infty)$	$r_{\text{osco}} \in (7.5, +\infty)$ $-$
$f(R)$ Kerr	$R_0 \in (0, 1.45 \times 10^{-1})$ $R_0 \in (-0.13, 0)$	$r_{\text{isco}} \in (1.4545, 2)$ $r_{\text{isco}} \in (1, 1.4545)$	$r_{\text{osco}} \in (2, +\infty)$ $-$

(1999)] and Rezzolla *et al.* [Rezzolla, Zanotti, and Font (2003)], and also in Kerr–de Sitter and Kerr–anti-de Sitter performed by Stuchlík and Slaný [Stuchlík and Slaný (2004)].

supermassive black holes ($10^7 M_\odot \leq M \leq 10^9 M_\odot$), the values of α were in the range $\alpha \in (0.03, 2.47)$. The lower limit for α was set in agreement with Moffat *et al.* [Moffat and Toth (2008)] that predicted such a value for globular clusters, while the upper limit for the parameter was calculated by fitting rotation curves of dwarf galaxies ($1.9 \times 10^9 M_\odot \leq 3.4 \times 10^{10} M_\odot$) by Brownstein and coworkers [Brownstein and Moffat (2006)].

Pérez *et al.* showed that for both stellar and supermassive black holes in Schwarzschild and Kerr STVG spacetimes, the innermost stable circular orbit (ISCO) is always larger than for the corresponding spacetimes in GR; this occurs for all the values of the parameter α considered by the authors. The major difference with respect to GR occurs for supermassive Kerr black holes ($a = 0.99$): the ISCO increases up to 716 percent with respect to the value of the ISCO in GR for $\alpha = 2.45$.

Notice that formulas (2.44) and (2.45) were derived from the laws of conservation of rest mass, angular momentum, and energy. These expressions, however, do not hold in STVG. A neutral test particle in STVG spacetime is subjected to a gravitational Lorentz force whose vector potential in Boyer–Lindquist coordinates is:

$$\phi = \frac{-Qr}{\rho^2}(\mathbf{dt} - a \sin \theta^2 \mathbf{d\phi}). \tag{2.47}$$

Here, Q is the gravitational source charge of the vector field ϕ^μ. Expressions (2.44) and (2.45) are now redefined as:

$$\tilde{E} = -\frac{p_t}{m} + \frac{q}{m}\phi_t, \tag{2.48}$$

$$\tilde{L} = \frac{p_\phi}{m} + \frac{q}{m}\phi_\phi, \tag{2.49}$$

and the conservation laws for angular momentum and energy take the form:

$$(\tilde{L} - w)_{,r} + \frac{q}{m}\phi_{\phi,r} = f\left(\tilde{L} + \frac{q}{m}\phi_\phi\right), \tag{2.50}$$

$$(\tilde{E} - \Omega w)_{,r} + \frac{q}{m}\phi_{t,r} = f\left(\tilde{E} + \frac{q}{m}\phi_t\right), \tag{2.51}$$

where

$$f = 4\pi e^{\nu+\Phi+\mu} F/\dot{M}_0, \tag{2.52}$$

$$w = 2\pi e^{\nu+\Phi+\mu} W_\phi{}^r/\dot{M}_0. \tag{2.53}$$

F denotes the emitted flux, $W_\phi{}^r$ the torque per unit circumference, $e^{\nu+\Phi+\mu}$ is the square root of the metric determinant, and $\dot{M}_0$ the mass accretion rate.

Once the heat Q emitted by the disk is computed, the temperature profile can be obtained by means of the Stefan–Boltzmann's law[18]:

$$T(x) = z(x) \left(\frac{Q(x)}{\sigma} \right)^{1/4}. \tag{2.54}$$

In this equation, $z(x)$ gives the correction due to gravitational redshift, and σ denotes the Stefan–Boltzmann constant. Under the black body hypothesis, the emissivity per unit frequency I_ν of each element of area of the disk is described by the Planck function:

$$I_\nu(\nu, x) = \frac{2h\nu^3}{c^2 \left[e^{(h\nu/kT(x))} - 1 \right]}. \tag{2.55}$$

Finally, the total luminosity at frequency ν is:

$$L_\nu = \frac{4\pi h G^2 M^2 \nu^3}{c^6} \int_{x_{isco}}^{x_{out}} \frac{x \, dx}{\left[e^{(h\nu/kT(x))} - 1 \right]}. \tag{2.56}$$

We show in Fig. 2.1 the temperature and luminosity distributions for an accretion disk around an $f(R)$–Kerr black hole for negative values of the Ricci scalar [Pérez, Romero, and Perez Bergliaffa (2013)]. Pérez and coworkers adopted for the values of the relevant parameters, i.e., M mass, $\dot{M}$ accretion rate, and a angular momentum of the black hole, the best estimates available for the galactic black hole Cyg X-1 [Gou (2011); Orosz *et al.* (2011)]. For $R_0 < 0$, the temperature and luminosity of the disk are always higher than in GR. In particular, for $R_0 = -1.25 \times 10^{-1}$ the peak of the emission rises a factor 2 with respect to GR. In order to fit this $f(R)$–Kerr model

---

[18]It is assumed that the disk is optically thick in the z-direction, so that every element of area on its surface radiates as a black body.

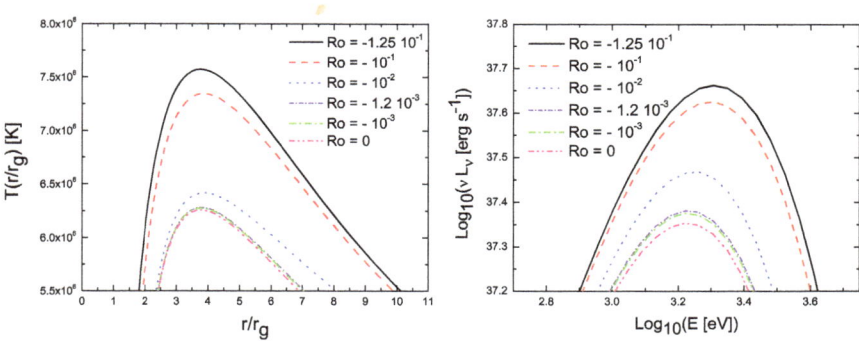

Fig. 2.1 $f(R)$–Kerr black hole of angular momentum $a = 0.99$ for $R_0 < 0$. From Pérez *et al.* [Pérez, Romero, and Perez Bergliaffa (2013)]. Left figure: Temperature as a function of the radial coordinate. Right figure: Luminosity as a function of the energy. Reproduced by permission of the authors.

with current observations of Cygnus X-1 in the soft state, curvature values below 1.2×10^{-3} have to be rule out. Accretion disk models for positive values of the Ricci scalar in the range $R_0 \in (0, 6.67 \times 10^{-4}]$ present no relevant differences compared to accretion disks in GR.

According to Pérez *et al.* [Pérez, Armengol, and Romero (2017)], accretion disks around both Schwarzschild and Kerr black holes in STVG are colder and underluminuous in comparison with thin relativistic accretion disks in GR. The greatest differences in temperature and luminosity were found for accretion disks around supermassive Kerr–STVG black holes, as depicted in Fig. 2.2. For instance, if $\alpha = 2.45$ the temperature decreases up to 12.30 percent, and the peak of the luminosity is about 4 percent lower than in GR.

To sum up, we have seen that investigations of accretion processes around black holes shed light on the behavior of gravity under extreme conditions. In particular, by comparing the spectral energy distributions predicted by modified theories of gravitation with current astronomical observations, more restrictive limits on the values of the free parameters of these theories can be set. For $f(R)$–Schwarzschild and $f(R)$–Kerr black holes, thin stable accretion disks are possible for $R_0 \in (-\infty, 10^{-6}]$ and $R_0 \in [-1.2 \times 10^{-3}, 6.67 \times 10^{-4}]$, respectively. In the case of Schwarzschild and

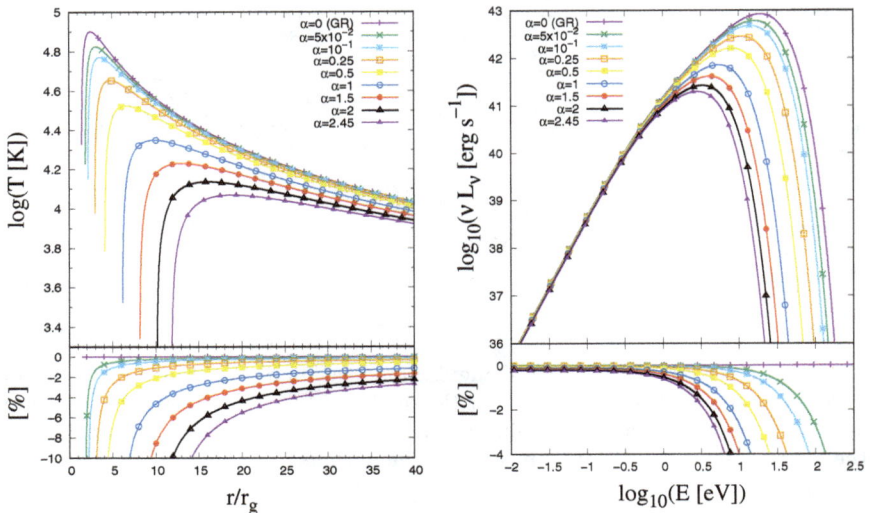

Fig. 2.2   Supermassive Kerr–STVG black hole of spin $\tilde{a} = 0.99$. From Pérez *et al.* [Pérez, Armengol, and Romero (2017)]. Left figure: On top, temperature as a function of the radial coordinate; bottom, residual plot of the temperature as a function of the radial coordinate. Right figure: On top, luminosity as a function of the energy; bottom, residual plot of the luminosity as a function of the energy. Reproduced by permission of the authors.

Kerr–STVG black holes, accretion disks can exist for stellar sources if $0 < \alpha < 0.1$, while for supermassive sources $0.03 < \alpha < 2.47$.

It should be mentioned that other features of accretion flows in $f(R)$-gravity were studied by some authors. Ahmed and coworkers [Ahmed (2016); Azreg-Aïnou (2017)] analyzed some aspects of the Michel-type accretion onto static spherically symmetric black holes in $f(R)$-gravity. Effects of radial and angular pressure gradients on thick accretion disks in $f(R)$–Schwarzschild geometry with constant Ricci curvature were investigated by Alipour *et al.* [Alipour, Khesali, and Nozari (2016)].

Bhattacharjee and collaborators [Bhattacharjee, Das, and Mahajan (2015)] researched additional sources of vorticity in accreted material by black holes, others than the classical baroclinic instability. In particular, they analyzed how the spacetime geometry contributes to the vorticity generation in accretion disk plasma, both in GR and $f(R)$-gravity with constant Ricci scalar. They found, as expected, that vorticity generation is more effective in Kerr than Schwarzschild spacetime, and also stronger in $f(R)$-gravity. The efficiency of this mechanism increases in the regions where strong gravity is dominant. Interestingly, the formalism developed by Bhattacharjee *et al.* [Bhattacharjee, Das, and Mahajan (2015)] can be extended to multifluid species, and might provide a novel mechanism for angular momentum transport.

2.6. Effects on Jets

The first observational evidence of the existence of astrophysical jets was due to Herber Curtis of Lick Observatory. In 1918, he observed in M87, a supergiant elliptical galaxy in the constellation Virgo, "a curious straight ray...apparently connected with the nucleus by a thin line of matter". Nowadays, we know that such a "straight ray" is a collimated flow of particles and electromagnetic fields ejected by the supermassive black hole that lies at the core of the galaxy. The highly collimated relativistic jet extends at least 1.5 kiloparsecs from the nucleus of M87 well into the intergalactic medium.

Jets are observed in a plethora of astrophysical systems, from protostars to active galactic nuclei. There seems to be some key ingredients necessary for the formation of relativistic jets: accretion onto a spinning compact object, and the presence of a large-scale magnetic field.

It is though that the launching region of relativistic jets in active galactic nuclei is near the event horizon of the supermassive black hole, where the strong gravity effects are important. Thus, it is reasonable to expect that any deviations from GR should manifest on such scales. These effects on relativistic jets were recently investigated by Lopez Armengol and Romero [Lopez Armengol and Romero (2017b)] in STVG. The aim of these authors was to compare GR and STVG close to the gravitational source and study the differences between the two theories on short scales.[19]

---

[19]Up to the moment this chapter was written, no works on relativistic jets in $f(R)$-gravity were found.

The method used by the authors was to calculate the trajectories of massive particles in Kerr–STVG spacetime. The black hole parameters M mass, and a angular momentum were adopted from the observational estimates for the supermassive black hole in M87 reported by Gebhardt *et al.* [Gebhardt *et al.* (2011)], and Li *et al.* [Li *et al.* (2009)].

The results of Lopez Armengol and Romero can be divided in two main parts. First, they studied the azimuthal orbits of massive particles set with initial position $r_0 = 140G_N M/c^2$, $\theta_0 = 0.18$, $\phi_0 = 0$ and initial Lorentz factor $\gamma = 2$. The values of the parameters were taken from the observational results by Mertens *et al.* [Mertens *et al.* (2016)]. Two different cases for the ejection angles were taken into account: $\theta_{ej}^A = 0$ and $\theta_{ej}^B = 0.3$.

The value of the free parameter α of the theory is related to the parameter M_0 by the formula:

$$\alpha = \sqrt{\frac{M_0}{M}}, \tag{2.57}$$

where M is the mass of the gravitational source. On the other hand, the parameter κ that appears in the equation of motion for a test particle:

$$\left(\frac{d^2 x^\mu}{d\tau^2} + \Gamma^\mu_{\alpha\beta} \frac{dx^\alpha}{d\tau} \frac{dx^\beta}{d\tau} \right) = \kappa B^\mu{}_\nu \frac{dx^\nu}{d\tau}, \tag{2.58}$$

is also linked with the parameter α according to the following expression postulated by Moffat [Moffat (2015)]:

$$\kappa = \sqrt{\alpha G_N}. \tag{2.59}$$

The calculation of the orbits was made for two runs of the parameters. In the first run, M_0 was fixed[20] ($M_0 = 10^{11} M_\odot$), and the parameter κ was sample: $\kappa_1 = 10^2 \sqrt{\alpha G_N}$, $\kappa_2 = 10^3 \sqrt{\alpha G_N}$, $\kappa_3 = 10^4 \sqrt{\alpha G_N}$.

---

[20]Lopez Armengol and Romero [Lopez Armengol and Romero (2017b)] established an upper limit for the value of M_0 using: 1) the observational estimates for the radius of M87* (supermassive black hole at the center of the elliptical galaxy M87) of $\approx 8G_N M/c^2$ by Broderick *et al.* [Broderick *et al.* (2015)], and 2) the formula for the event horizons of a Kerr–STVG black hole (see Sec. 2.4.2, Eq. (2.41)).

Significant deviations from GR were found: the angular velocity ω_ϕ as a function of the z coordinate is enhanced or diminished by the gravito-magnetic forces depending on the initial ejection angle. Also, the augmented gravito-electric repulsion due to the growing values of κ increases the local Lorentz factor γ of the particles with time. The latter implies that in the strong field regime of STVG, particles gravitationally accelerate.

The values of the local Lorentz factor γ, however, cannot be arbitrarily large. The highest value of γ observationally estimated for M87 corresponds to the spine of the jet and is $\gamma \approx 10$ [Mertens *et al.* (2016)]. Thus, the upper limit imposed to κ is [Lopez Armengol and Romero (2017b)]:

$$\kappa \leq 10^2 \sqrt{\alpha\, G_{\mathrm{N}}}. \tag{2.60}$$

In the second run, Lopez Armengol and Romero took Moffat's weak field limit prescription $\kappa = \sqrt{\alpha G_{\mathrm{N}}}$, and set $\theta_{\mathrm{ej}}^{\mathrm{A}} = 0$. In this case, the parameter M_0 was modified: $M_0 = 10^{10} M_\odot$, $M_0 = 10^{11} M_\odot$, and $M_0 = 10^{12} M_\odot$. Though the latter value violates the condition $M_0 \leq 10^{11} M_\odot$ computed by [Lopez Armengol and Romero (2017b)] taking into account observational constraints for M87*, it was included for consistency checks.

Contrary to the first run, the effects of the Lorentz-like forces on trajectories became insignificant. This could be due to the small value of κ. The increment in the value of α, through the parameter M_0, led to a larger decrement of the particle velocity in contrast to GR.

Finally, the authors compared observational results on the formation zone of the jet in M87 with predictions of STVG. The values selected to model the jet were $r_0 = 5R_{\mathrm{S}}$, $M_0 = 10^{10} M_\odot$, $\kappa_1 = 10 \sqrt{\alpha G_{\mathrm{N}}}$, $\kappa_2 = 10^2 \sqrt{\alpha G_{\mathrm{N}}}$, $\kappa_3 = 10^3 \sqrt{\alpha G_{\mathrm{N}}}$; a wide ejection angle was assumed in accordance to the Blandford–Payne mechanism for jet launching [Blandford and Payne (1982); Spruit (2010)]. In Fig. 2.3, the x-z trajectories for different values of κ are plotted. The filled region corresponds to the jet according to the parametrization by Mertens *et al.* [Mertens *et al.* (2016)]. Notably, the larger the values of κ, the higher is the collimation of the jet.

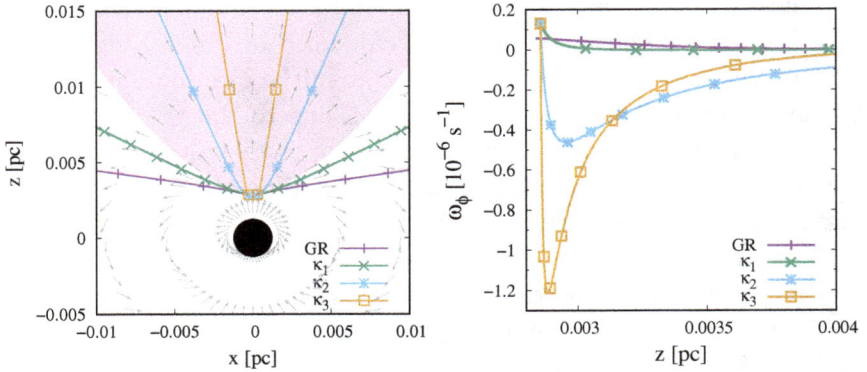

Fig. 2.3   The gravito-magnetic field in STVG severely changes the trajectories of particles giving rise to effects that are absent in GR. Left: Projection in the x-z plane of the trajectories of massive particles at the base of jet, for different values of κ. The filled region represents the jet as modeled by Mertens *et al.* [Mertens *et al.* (2016)]. Right: Angular velocity as a function of the z coordinate for particles ejected with a wide angle at the launching region of the jet. Reprinted by permission from Springer Customer Service Centre GmbH: Springer Nature, *Astrophysics and Space Science*, Lopez Armengol, F.G. & Romero, G.E. (*Astrophys. Space Sci.* (2017) 362: 214. https://doi.org/10.1007/s10509-017-3197-6).

The plot of the angular velocity ω_ϕ as a function of z for different values of κ is shown in Fig. 2.3 at the right. At the base of the jet, the gravito-magnetic force leads to a counter rotation in ϕ. Since the field lines rotate along the trajectory, from certain value of z, the sign of the gravito-magnetic force changes and ω_ϕ starts to increase. The latter result is particularly interesting in the light of the recent observations of the jet of M87 [Mertens *et al.* (2016)] in which, for the first time, rotation has been directly measured. The jet component closer to the launching region rotates clockwise while further from M87, the jet rotates counterclockwise [Mertens *et al.* (2016)]. The calculations of Lopez Armengol and Romero for the scale where rotation changes sign and the order of magnitude of the angular velocity ω_ϕ are in agreement with the estimations by Mertens and coworkers [Mertens *et al.* (2016)].

In conclusion, we have seen that STVG offers an alternative mechanism for jet formation of purely gravitational origin that is compatible with current observational data.

2.7. Gravitational Waves

In this section we will focus on the implications for $f(R)$-gravity and STVG of the recent detection of gravitational waves by the LIGO Scientific Collaboration and Virgo Collaboration.[21]

Capozziello and collaborators [Capozziello, Corda, and de Laurentis (2008)] were the first to generalize some previous results on gravitational waves in $f(R)$-gravity. They showed that there are only three propagating degrees of freedom in such theory: the massless plus and cross polarizations which are the same as in GR, and a third massive scalar mode given by a mixed of longitudinal and transverse polarizations (see also Liang *et al.* [Liang *et al.* (2017)]).

The signal that a gravitational wave detector would identify if gravitational waves had such additional polarization modes was computed by Corda [Corda (2018)]. He provided expressions for the frequency and angular dependent response function of a gravitational wave interferometric detector in the presence of a third massive mode (see Eq. (59) in the paper by Corda [Corda (2018)]). The formulas given by Corda are of particular importance since they allow to discriminate between massless and massive modes in scalar–tensor gravity and $f(R)$-gravity.

Current ground based gravitational wave interferometers do not have the sensitivity to detect the directions in which the instrument is oscillating. For the transient waves already observed, a study of the gravitational wave polarization cannot yet be done with the LIGO–Virgo network, since at least five non-co-oriented arm antennas are required to break the degeneracies between the five distinguishable modes of a generic metric theory of gravity [Corda (2018); Abbott *et al.* (2018)]. Nonetheless, studies of the polarization content of the signal have already been conducted. This was the case for the gravitational wave event GW170814 produced by the merger of two stellar mass black holes observed with both the LIGO and

---

[21]The reader interested in more theoretical aspect of gravitational waves in $f(R)$-gravity is referred to [Sotiriou and Faraoni (2010); Capozziello and Faraoni (2011)]. See also Capozziello and Bajardi (2019).

Virgo detectors [Abbott *et al.* (2017b)]; the analysis of the data strongly favors pure tensor polarization of gravitational waves, over pure scalar or pure vector polarizations.

Theory independent polarization measurements, however, can be done with the current instruments if the detectors are exposed to sufficiently long signals. This occurs for pulsars which are expected to emit continuous gravitational waves. At the beginning of 2018, the LIGO Scientific Collaboration and Virgo Collaboration [Abbott *et al.* (2018)] presented the results of the first directed search of non-tensorial gravitational waves. The investigation focused on 200 known pulsars using data from aLIGO's first observation run; no assumption was made about the polarization modes of the gravitational waves. The data showed no evidence for the emission of gravitational signals of tensorial or non-tensorial polarization from any of the pulsars studied. They also obtained upper limits for the strain of the scalar and vector modes (1.5×10^{-26} at 95 percent credibility) that can in turn be used to constraint alternative theories of gravity. Notice that an important assumption of the work was that the gravitational wave emission frequency f_{GW} was twice of the rotational frequency f_{rot} of the source, that is $f_{GW} = 2f_{rot}$. This hypothesis follows the most favored emission model in GR. Such condition will be relaxed in future investigations.

The Advanced LIGO–Virgo network has also carried out a search for a generically polarized stochastic background of gravitational waves [Abbott *et al.* (2018a)] using data from Advanced LIGO's O1 observing run.[22] No evidence was found for the presence of a background of gravitational waves, and of any polarization. However, direct upper limits were established on the contributions of vector and scalar modes to the stochastic background.

A key point recently clarified by Corda [Corda (2018)] is that the constraints on the graviton mass derived by the LIGO Scientific

---

[22]The gravitational wave background is a random gravitational wave signal that is supposed to be generated by distant compact binary mergers, core-collapse supernovae, and rapidly rotating neutron stars; a background of cosmological origin may also be present (see for instance [Caprini and Figueroa (2018)]).

Collaboration and the Virgo Collaboration [Abbott *et al.* (2017a)] are not on some extra polarization mode of the gravitational waves but on the tensor modes. In fact, the upper limit on the graviton mass is derived assuming that gravitons disperse in vacuum as massive particles; the lack of dispersion in the gravitational waves sets a lower limit on the Compton wavelength $\lambda_g > 1.6 \times 10^{13}$ km, or for the graviton mass $m_g < 7.7. \times 10^{-23}$ eV/c^2 [Abbott *et al.* (2017a)].

Some authors have used the constraint on the graviton mass mentioned above to bound free parameters in different models of $f(R)$-gravity. For instance, Vainio and Vilja [Vainio and Vilja (2018)] constrained the parameter μ in the Hu–Sawicki model [Hu and Sawicki (2007)] (taking $n \approx 1$), and also the parameter λ in Starobinsky model [Starobinsky (2007)]. A similar procedure was employed by Lee [Lee (2018)] to obtain limits on the parameter M_0 of a generally constructed $f(R)$ model determined from cosmological observations [Lee (2019)].

On August 17th, 2017, it was observed for the first time a neutron star merger in gravitational waves [Abbott *et al.* (2017c)] (GW170817). The electromagnetic counterpart [Abbott *et al.* (2017e,d)], GRB170817A, was detected 1.7 s after GW170817. The observed delay between both events was used to restrict the difference between the speed of gravitational waves and the speed of light:

$$\left|\frac{c_{\text{GW}}}{c} - 1\right| < 5 \times 10^{-16}, \tag{2.61}$$

where c_{GW} denotes the speed of the gravitational waves, and c stands for the speed of light. This result was used by some authors to show that a large class of alternative theories of gravitation are on the verge of being completely discarded (see, for instance, [Ezquiaga and Zumalacárregui (2017); Pardo *et al.* (2018)]). Some theories, however, have survived such stringent test. Nojiri and Odintsov [Nojiri and Odintsov (2018)] explicitly showed that in $f(R)$-gravity the propagating light speed c is identical to the propagating speed c_{GW} of the gravitational waves. There is some difference, however, in the gravitational wave propagation phase with respect to the light one; the authors suggest future observations that could detect the shift of the phase.

A novel generalized framework for testing the nature of gravity using gravitational wave propagation has just been introduced by Nishizawa [Nishizawa (2018)]. The method employed was to analytically solve the gravitational wave propagation equation in an effective field theory for dark energy and get a WKB solution. The gravitational waveform obtained contains functions of time that characterize modified amplitude damping, modified propagation speed, non-zero graviton mass, and also a source term for the gravitational waves. The author also provided specific expressions of these general functions in the context of various alternative theories of gravitation. In a second paper, Arai and Nishizawa [Arai and Nishizawa (2018)] applied this generalized framework to test all possible models of the Horndeski theory. Using the data from the simultaneous detection of GW170817 and GRB170817A, some models within the theory were excluded while quintessence, non-linear kinematic theory, and $f(R)$-gravity were favored.

What was the fate of STVG after the six gravitational wave detections by the LIGO–Virgo Collaboration?

As we already mentioned, the detection of gravitational waves produced by the merger of a binary system of neutron stars, and the subsequent observation of a short gamma-ray burst (GRB170817A), provided invaluable data to test GR and also alternative theories of gravitation.

A classical test of GR is the Shapiro time delay, or gravitational time delay [Shapiro (1964)]. GR predicts that the amount of time it takes an electromagnetic signal to travel to a target and return is longer if a massive object is close in its path. The time delay is caused by spacetime dilatation, which increases the path length. In GR, the Shapiro time delay is the same for gravitational waves and photons since both travel on null geodesics. This may not be the case in other theories of gravity, in particular in those modified theories of gravity which dispense of dark matter and reproduce the modified

Newtonian dynamics (for instance galaxy rotation curves) in the non-relativistic limit, the so-called *Dark Matter Emulators* [Kahya and Woodard (2007)].

Theories of gravity dubbed "Dark Matter Emulators" were defined explicitly as [Kahya and Woodard (2007); Boran *et al.* (2018)]:

(1) Ordinary matter couples to the metric $\tilde{g}_{\mu\nu}$ ($\tilde{g}$ denotes the "disformally transformed metric") that would be produced by GR with dark matter, and

(2) Gravitational waves couple to the metric $g_{\mu\nu}$ produced by GR without dark matter.

From the above definitions it follows that in Dark Matter Emulators, photons suffer an additional Shapiro time delay due to the dark matter needed if GR was correct. Boran and collaborators [Boran *et al.* (2018)] estimated that the Shapiro time delay due to the gravitational potential of the total dark matter distribution along the line of sight from NGC 4993 to the Earth is of the order of 400 days. Since the electromagnetic detection of GW170817 was almost immediate while Dark Matter Emulators predict delays over a year, Boran *et al.* concluded that these theories were rule out.

In the realm of Dark Matter Emulators, Boran *et al.* included STVG, which we have extensively discussed in this chapter. Green, Moffat and Toth [Green, Moffat, and Toth (2018)] have recently demonstrated that STVG do not belong to such class of theories.

Green *et al.* [Green, Moffat, and Toth (2018)] explicitly showed that in STVG:

- Gravitational waves move at the speed of light as photons do.
  STVG is constructed on these gravitational fields: the metric $g_{\mu\nu}$ (a spin 2 massless graviton), a scalar field $G = G_{\mathrm{N}}(1 + \alpha)$ (a spin 0 massless graviton), a vector field ϕ_μ (a spin 1 repulsive massive graviton), and a spin 0 scalar field denoted μ that is the mass of the vector field ϕ_μ. The mass of μ, represented by m_ϕ, was estimated from fits to galaxy rotation curves and clusters without

dark matter [Moffat and Rahvar (2013, 2014)], and it is $m_\phi = 2.8 \times 10^{-28}$ eV, which is of the order of the experimental bound on the photon mass. Since the vector particle mass is extremely small, the three gravitons that correspond to $g_{\mu\nu}$, $G = G_N(1 + \alpha)$, and ϕ_μ move at the speed of light.

- The weak equivalence principle is satisfied.
  The equation of motion of a massive test particle in STVG is [Moffat (2006)]:

$$m \left(\frac{d^2 x^\mu}{d\tau^2} + \Gamma^\mu_{\alpha\beta} \frac{dx^\alpha}{d\tau} \frac{dx^\beta}{d\tau} \right) = q_g B^\mu{}_\nu \frac{dx^\nu}{d\tau}, \qquad (2.62)$$

where $u^\mu = dx^\mu/ds$ and $q_g = \kappa m$ is the gravitational charge of the test particle [Moffat (2006)] (see also Sec. 2.6). We can cancel the mass on both sides of the equation above thus showing that all particles (bodies) move on spacetime independently of their constitution.

We see, then, that STVG is still standing after the observations of GW170817 and GRB170817A. Shall it be the case in the future? We do not know. What we do know is that further research is needed on the nature of gravitational waves in STVG. To date, it has been only derived the linearized field equations and gravitational energy flux in the weak field regime [Moffat (2016)]. In the case of well separated binary pulsars the tensor radiated power reduces to the corresponding result in GR [Moffat (2016)] in agreement with observations of binary pulsars.[23]

The process of the merger of a system of binary black holes can be described in three phases: the inspiral, merger, and ringdown. The final stage, the ringdown, is when the resulting black hole horizon settles down through damped oscillations which can in turn be characterized by the quasi-normal modes (QNMs).

In the first gravitational wave event detected [Abbott *et al.* (2016a)], GW150914, it was possible to obtain the frequency and

---

[23]Moffat obtained in STVG constraints of some parameters related to the spins of spiralling binary black holes. He also showed that STVG predicts a misaligning of the black holes that are in coalesce [Moffat (2017)].

damping time of the QNMs. The data were used to derive the mass M and spin a of the black hole [Abbott *et al.* (2016b)]. If in the future additional QNMs are detected, i.e., first overtone, it may be possible to distinguish deviations from GR. For instance, the frequency of QNMs of Schwarzschild black holes in STVG is greater than in GR as the value of the free parameter α of the theory increases [Manfredi, Mureika, and Moffat (2018)] (see also Wei *et al.* [Wei and Liu (2018)] for some related results for Kerr black holes in STVG). In $f(R)$-gravity, Bhattacharyya and Shankaranarayanan [Bhattacharyya and Shankaranarayanan (2017)] showed that vector and scalar perturbations in $f(R)$ black hole spacetimes do not emit the same amount of gravitational wave energy. Indeed, one of the most relevant differences between GR and $f(R)$-gravity is the existence, in the latter, of dipole and monopole radiation. This is a consequence of the additional degree of freedom of the theory represented by the scalar field.[24]

2.8. Singularities and Beyond

Singularities appear under several circumstances in GR. Gravitational collapse, in quite general conditions, results in singularities [Penrose (1965)]. Cosmological models of the expanding universe are also singular for normal matter contents and GR [Hawking and Penrose (1970)]. All these singularities are pathological features, since they imply a breakdown of the corresponding models and their predictive power [Romero (2013)]. Singularities are not entities of any kind, but a sign of the incompleteness of the underlying theory. They are not expected to show up in a consistent theory of quantum gravity. But also classical theories can be free of singularities if they assume a more complex structure for spacetime.

---

[24]The computation of the gravitation radiation has been done for different choices of the $f(R)$ function; see for instance [Näf and Jetzer (2011); Bergliaffa and Nunes (2011); de La Cruz-Dombriz, Dobado, and Maroto (2011)].

Non-singular black holes have been discussed in the framework of $f(R)$ theories by Olmo and Rubiera-Garcia [Olmo, Rubiera-Garcia and Sanchez-Puente (2015b, 2016)]. They have found that in quadratic gravity and for anisotropic fluids a region with non-trivial topology (a so-called wormhole) replaces the singularity. The resulting spacetime is geodesically complete although curvature divergence exists in the wormhole throat. With more generality Bejarano *et al.* [Bejarano, Olmo, and Rubiera-Garcia (2017)] have investigated the relation between energy density, curvature invariants, and geodesic completeness in such spacetimes. They have explored how the anisotropic fluid helps to modify the innermost geometry of the black hole. A variety of configurations with and without wormholes have been found, as well as solutions with de Sitter interiors, solutions that mimic non-linear models of electrodynamics coupled to GR, and configurations with up to four horizons. The fact that several of these models show divergences in the curvature scalar but nevertheless remain geodesically complete shows that simplistic analyses based only on the behavior of curvature invariants can be misleading. A remarkable feature is that the anisotropic fluid has a stress–energy tensor satisfying the energy conditions, which are usually violated to generate regular black holes in GR (see, for instance, the work of Pérez *et al.* [Pérez, Romero, and Perez Bergliaffa (2014)]).

Regular black holes can be also obtained in STVG gravity. Moffat has discussed both static and rotating solutions [Moffat (2015, 2018)]. In the case of the Schwarzschild-like black hole the solution reads:

$$ds^2 = \left(1 - \frac{2GMr^2}{(r^2 + \alpha G_N GM^2)^{3/2}} + \frac{\alpha G_N GM^2 r^2}{(r^2 + \alpha G_N GM^2)^2}\right) dt^2$$
$$- \left(1 - \frac{2GMr^2}{(r^2 + \alpha G_N GM^2)^{3/2}} + \frac{\alpha G_N GM^2 r^2}{(r^2 + \alpha G_N GM^2)^2}\right)^{-1} dr^2$$
$$- r^2 d\Omega^2. \tag{2.63}$$

This metric is regular at $r = 0$ and asymptotically flat as $r \to \infty$. For large r it reproduces the Schwarzschild metric. For small r the metric behaves as

$$ds^2 = \left(1 - \frac{1}{3}\Lambda r^2\right) dt^2 - \left(1 - \frac{1}{3}\Lambda r^2\right)^{-1} dr^2 - r^2 d\Omega^2, \qquad (2.64)$$

where the effective cosmological constant Λ is given by

$$\Lambda = \frac{3}{G_N^2 M^2} \left(\frac{\alpha^{1/2} - 2}{\alpha^{3/2}(1 + \alpha)}\right). \qquad (2.65)$$

So the interior material of the regular STVG black hole satisfies the vacuum equation of state $p = -\rho$, where p and $\rho = \rho_{\text{vac}}$ are the pressure and the vacuum density, respectively. The effective cosmological constant Λ can be positive or negative depending on the magnitude of α, so that the interior of the black hole is described by either a de Sitter or anti-de Sitter spacetime. Shadows of these black holes are expected to be different from GR black holes of the same mass. Hence, Event Horizon Telescope observations can be used to test the theory [Moffat (2015)].

Another interesting feature of STVG is that it allows for wormhole solutions that do not require exotic matter. If wormholes exist in nature they might be found by gravitational lensing, as pointed out by Safonova *et al.* [Safonova, Torres, and Romero (2002)]. Wormhole lensing events and wormhole shadows are expected to differ in STVG with respect to GR.

One interesting feature of $f(R)$-gravity when applied to the whole universe is that early time cosmology can be adjusted in such a way as to guarantee the existence of homogeneous and isotropic models that avoid the Big Bang singularity. In such models the Big Bang singularity can be replaced by a cosmic bounce without violating any energy condition. The bounce is possible even for pressureless dust [Barragán, Olmo, and Sanchis-Alepuz (2009); Paul, Chakrabarty, and Bhattacharya (2014); Bhattacharya and Chakrabarty (2016)]. Such models can be tested through measurements of the modes B of the CMB polarization by instruments such as the forthcoming QUBIC [Aumont *et al.* (2016)]. Specifically, it has been shown that there are distinctive (oscillatory) signals on the primordial gravitational wave spectrum for very low frequencies in $f(R)$-gravity; such signals correspond to modes that are currently entering the horizon [Bouhmadi-Lopez, Morais, and Henriques (2013)].

Detailed studies regarding the potential role of STVG in producing cosmological bounces are still missing, likely because of the complexity of the theory. Recently, Jamali *et al.* [Jamali, Roshan, and Amendola (2018)] have shown that extra fields in STVG cannot provide a late time accelerated expansion. Furthermore, they have solved the non-linear field equations numerically and calculated the angular size of the sound horizon with results outside the current observational bounds. However, further research is necessary since the models analyzed so far are rather simple, without self-interactions.

Just as in the last century the universe was investigated through the whole electromagnetic spectrum, the 21st century is the starting of a new time in which gravitational waves will be detected over a wide range of frequencies. A new generation of gravitational wave detectors in space and underground, eLISA, KAGRA, and the Einstein Telescope are planned to start working in the next decades. Pulsar timing monitoring will also make possible the investigation of the exceedingly long wavelength perturbations in spacetime curvature caused by the merging of supermassive black holes [Arzoumanian *et al.* (2018); Ellis (2014)]. Alternative theories of gravity, and GR itself, will be put to the test to a level of detail that is impossible today. Novel aspects of the nature of gravity might be unveiled that perhaps none of the actual theories predict or foresee.

Acknowledgments

We thank S.E. Perez Bergliaffa, John Moffat, F. López Armengol, and L. Combi for many discussions on these topics. This work was supported by CONICET (PIP2014-0338) and by the Spanish Ministerio de Economía y Competitividad (MINECO) under grant AYA2013-47447-C3-1-P and AYA2016-76012-C3-1-P.

References

Abbott, B.P. *et. al.*, *Phys. Rev. Lett.* **116**, 061102 (2016a).
Abbott, B.P. *et. al.*, *Phys. Rev. Lett.* **116**, 221101 (2016b).
Abbott, B.P. *et. al.*, *Phys. Rev. Lett.* **116**, 241103 (2016c).
Abbott, B.P. *et. al.*, *Phys. Rev. Lett.* **118**, 221101 (2017a).
Abbott, B.P. *et. al.*, *Phys. Rev. Lett.* **119**, 141101 (2017b).
Abbott, B.P. *et. al.*, *Phys. Rev. Lett.* **119**, 161101 (2017c).
Abbott, B.P. *et. al.*, *Astrophys. J.* **848**, L12 (2017d).
Abbott, B.P. *et. al.*, *Astrophys. J.* **848**, L13 (2017e).
Abbott, B.P. *et. al.*, *Astrophys. J.* **851**, L35 (2017f).
Abbott, B.P. *et. al.*, *Phys. Rev. Lett.* **120** (3), 031104 (2018).
Abbott, B.P. *et. al.*, *Phys. Rev. Lett.* **120** (20), 201102 (2018).
Ahmed, A.K. *et al.*, *Eur. Phys. J. C* **76**, 280 (2016).
Alavirad, H. and Weller, J.M., *Astrophys. J.* **88** (12), 124034 (2013).
Alipour, N., Khesali, A.R., and Nozari, K., *Astrophys. Space Sci.* **361** (7), 240 (2016).
Amirabi, Z., Halilsoy, M., and Mazharimousavi, S.H., *Eur. Phys. J. C* **76**, 338 (2016).
Antoniadis, J. *et al.*, *Science* **340**, 448 (2013).
Aparicio Resco, M. *et al.*, *Physics of the Dark Universe* **13**, 147 (2016).
Arai, S. and Nishizawa, A., *Phys. Rev. D* **97** (10), 104038 (2018).
Arapoğlu, S., Deliduman, C., and Ekşi, K.Y., *J. Cosm. Astropart. Phys.* **7**, 020 (2011).
Arzoumanian, Z. *et al.*, *Astrophys. J.* **859**, 47 (2018).
Astashenok, A.V., Capozziello, S., and Odintsov, S.D., *Phys. Lett. B* **742**, 160 (2015).
Astashenok, A.V., Capozziello, S., and Odintsov, S.D., *J. Cosm. Astropart. Phys.* **1**, 001 (2015).
Astashenok, A.V., Odintsov, S.D., and de la Cruz-Dombriz, A., *Class. Quantum Grav.* **34**, 205008 (2017).
Aumont, J. *et al.*, arXiv:160904372A (2016).
Azreg-Aïnou, M., *Eur. Phys. J. C* **77**, 36 (2017).
Baade, W. and Zwicky, F., *Proc. Nat. Acad. Sci.* **20**, 254 (1934).
Babichev E., *et al.*, *Class. Quantum Gravity* **23**, 235014 (2016).
Bakirova, E. and Folomeev, V., *Gen. Relativ. Gravit.* **48**, 135 (2016).
Barausse, E. and Sotiriou, T.P., *Phys. Rev. Lett.* **101** (9), 099001 (2008).
Barausse, E., Sotiriou, T.P., and Miller, J.C., *Class. Quantum Grav.* **25** (6), 062001 (2008).
Barausse, E., Sotiriou, T.P., and Miller, J.C., *Class. Quantum Grav.* **25** (10), 105008 (2008).
Barragán, C., Olmo, G.J., and Sanchis-Alepuz, H., *Phys. Rev. D* **80** (2), 024016 (2009).
Bazeia, D. *et al.*, *Phys. Rev. D* **90** (4), 044011 (2014).

Bejarano, C., Olmo, G.J., and Rubiera-Garcia, D., *Phys. Rev. D*, **95** (6), 064043 (2017).

Bergliaffa, S.E.P. and Nunes, Y.E.C.D.O., *Phys. Rev. D* **84** (8), 084005 (2011).

Bhattacharjee, C., Das, R., and Mahajan, S.M., *Phys. Rev. D* **91**, 123005 (2015).

Bhattacharya, K. and Chakrabarty, S., *J. Cosm. Astropart. Phys.* **2**, 030 (2016).

Bhattacharya, K. and Chakrabarty, S., *J. Cosm. Astropart. Phys.* **2**, 30 (2016).

Bhattacharyya, S. and Shankaranarayanan, S., *Phys. Rev. D* **96**, 064044 (2017).

Blandford, R.D. and Payne, D.G., *Mon. Not. Roy. Astron. Soc.* **199**, 883 (1982).

Blázquez-Salcedo, J.L. *et al.*, *Phys. Rev. D* **98** (10), 104047 (2018).

Bolton, C.T., *Nature* **240**, 124 (1972).

Boran, S. *et al.*, *Phys. Rev. D* **97** (4), 041501 (2018).

Bouhmadi-Lopez, M., Morais, J., and Henriques, A.B., arXiv:1302.2038B (2013).

Broderick, A.E. *et al.*, *Astrophys. J.* **805**, 179 (2015).

Brownstein, J.R. and Moffat, J.W., *Mon. Not. Roy. Astron. Soc.* **367**, 527 (2006).

Calzà, M., Rinaldi, M., and Sebastiani, L., *Eur. Phys. J. C* **78**, 178 (2018).

Camenzind, M., *Compact Objects in Astrophysics: White Dwarfs, Neutron Stars, and Black Holes* (Springer-Verlag, Berlin, 2007).

Cañate, P., Jaime, L.G., and Salgado, M., *Class. Quantum Grav.* **33** (15), 155005 (2016).

Cañate, P., *Class. Quantum Grav.* **35** (2), 025018 (2018).

Capozziello, S. and Bajardi, F., *Int. J. Mod. Phys. D* **28**, 1942002 (2019).

Capozziello, S., Corda, C., and de Laurentis, M.F., *Phys. Lett. B* **669**, 255 (2008).

Capozziello, S. and Faraoni, V., *Beyond Einstein Gravity* (Springer-Verlag, Berlin, 2011).

Capozziello, S. *et al.*, *Phys. Rev. D* **93** (2), 023501 (2016).

Caprini, C. and Figueroa, D.G., *Class. Quantum Grav.* **35** (16), 163001 (2018).

Carter, B., *Phys. Rev. Lett.* **26**, 331 (1971).

Carter, B., in *Black Hole Equilibrium States*, Eds. C. DeWitt and B.S. DeWitt (Gordon and Breach. Science Publishers, New York, 1973).

Cembranos, J.A.R., de la Cruz-Dombriz, A., and Jimeno Romero, P., arXiv:1109.4519C (2011).

Cheoun, M.-K. *et al.*, *J. Cosm. Astropart. Phys.* **10**, 021 (2013).

Cooney, A., Dedeo, S., and Psaltis, D., *Phys. Rev. D* **82** (6), 064033 (2010).

Corda, C., *Int. J. Mod. Phys. D* **27** (5), 1850060 (2018).

de La Cruz-Dombriz, A., Dobado, A., and Maroto, A.L., *Phys. Rev. D* **80** (12), 124011 (2009).

de La Cruz-Dombriz, A., Dobado, A., and Maroto, A.L., Erratum: *Phys. Rev. DPRVDAQ1550-7998* **80**, 124011 (2009); *Phys. Rev. D* **83** (2), 029903 (2011).

De Laurentis, M. *et al.*, *Phys. Rev. D* **97**, 104024 (2017).

Deliduman, C., Ekşi, K.Y., and Keleç, V., *J. Cosm. Astropart. Phys.* **5**, 036 (2012).

Demorest, P.B. *et al.*, *Nature* **467**, 1081 (2010).

Douchin, F. and Haensel, P., *Astron. Astrophys.* **380**, 151 (2001).

Einstein, A., *Annalen der Physik.* **354**, 769 (1916).

Ellis, J., *APS Meeting Abstracts*, C15.008 (2014).

Ezquiaga, J.M. and Zumalacárregui, M., *Phys. Rev. Lett.* **119** (25), 251304 (2017).

Faraoni, V., *Phys. Rev. D* **81** (4), 044002 (2010).

Folomeev, V., *Phys. Rev. D* **97** (12), 124009 (2018).

Gao, C. and Shen, Y.-G., *Gen. Relativ. Gravit.* **48**, 131 (2016).

Gebhardt, K. *et al.*, *Astrophys. J.* **729**, 119 (2011).

Giacconi, R. *et al.*, *Phys. Rev. Lett.* **9**, 439 (1962).

Giddings, S.B., arXiv:hep-th/9508151 (1995).

Goriely, S., Chamel, N., and Pearson, J.M., *Phys. Rev. C.* **82** (3), 035804 (2010).

Gou, L. *et al.*, *Astrophys. J.* **742**, 85 (2011).

Green, M.A., Moffat, J.W., and Toth, V.T., *Phys. Lett. B* **780**, 300 (2018).

Habib Mazharimousavi, S., Halilsoy, M., and Tahamtan, T., *Eur. Phys. J. C* **72**, 1851 (2012).

Habib Mazharimousavi, S., Halilsoy, M., and Tahamtan, T., *Eur. Phys. J. C* **72**, 1958 (2012).

Habib Mazharimousavi, S., Kerachian, M., and Halilsoy, M., *Int. J. Mod. Phys. D* **22**, 1350057 (2013).

Harko, T., Kovács, Z., and Lobo, F.S.N., *Phys. Rev. D* **79** (6), 064001 (2009).

Hawking, S.W. and Penrose, R., *Proc. Roy. Soc. Lond. Ser. A* **314**, 529 (1970).

Hawking, S.W., *Commun. Math. Phys.* **25**, 152 (1972).

Hazard, C., Mackey, M.B., and Shimmins, A.J., *Nature* **197**, 1037 (1963).

Hendi, S.H. and Momeni, D., *Eur. Phys. J. C* **71**, 1823 (2011).

Hendi, S.H., Panah, B.E., and Mousavi, S.M., *Gen. Relativ. Gravit.* **44**, 835 (2012).

Hewish, A. *et al.*, *Nature* **217**, 709 (1968).

Hu, W. and Sawicki, I., *Phys. Rev. D* **76** (2007).

Israel, W., *Phys. Rev.* **164**, 1776 (1967).

Israel, W., *Commun. Math. Phys.* **8**, 245 (1968).

Jamali, S., Roshan, M., and Amendola, L., *J. Cosm. Astropart. Phys.* **1**, 48 (2018).

Kahya, E.O. and Woodard, R.P., *Phys. Lett. B* **652**, 213 (2007).

Kainulainen, K., Reijonen, V., and Sunhede, D., *Phys. Rev. D* **76** (4), 043503 (2007).

Kiziltan, B. *et al. Astrophys. J.* **778**, 66 (2013).

Kleihaus, B. *et al.*, *Phys. Rev. D* **93** (6), 064077 (2016).

Lee, S., *Physics of the Dark Universe* **25**, 100305 (2019).

Lee, S., *Eur. Phys. J. C* **78**, 449 (2018).

Li, Y.R. *et al.*, *Astrophys. J.* **699**, 513 (2009).

Liang, D. *et al.*, *Phys. Rev. D* **95**, 104034 (2017).

Liu, T., Zhang, X., and Zhao, W., *Phys. Lett. B* **777**, 286 (2018).

Lopez Armengol, F.G. and Romero, G.E., *Gen. Relativ. Gravit.* **49**, 27 (2017).

Lopez Armengol, F.G. and Romero, G.E., *Astrophys. Space Sci.* **362**, 214 (2017).

Manfredi, L., Mureika, J., and Moffat, J., *Phys. Lett. B* **779**, 492 (2018).

Maselli, A. *et al.*, *Phys. Rev. D.* **93** (12), 124056 (2016).

Mertens, F., *Astron. Astrophys.* **595**, A54 (2016).

Moffat, J.W., *J. Cosm. Astropart. Phys.* **3**, 004 (2006).

Moffat, J.W. and Toth, V.T, *Astrophys. J.* **680**, 1158 (2008).

Moffat, J.W. and Rahvar, S., *Mon. Not. Roy. Astron. Soc.* **436**, 1439 (2013).

Moffat, J.W. and Rahvar, S., *Mon. Not. Roy. Astron. Soc.* **441**, 3724 (2014).

Moffat, J.W., *Eur. Phys. J. C* **75**, 175 (2015).

Moffat, J.W., arXiv:151007037M (2015).

Moffat, J.W. and Toth, V.T, *Phys. Rev. D* **4**, 043004 (2015).

Moffat, J.W., *Phys. Lett. B* **763**, 427 (2016).

Moffat, J.W., arXiv:170605035M (2017).

Moffat, J.W., arXiv:180601903M (2018).

Moon, T., Myung, Y.S., and Son, E.J., *Gen. Relativ. Gravit.* **43**, 3079 (2011).

Moon, T., Myung, Y.S., and Son, E.J., *Eur. Phys. J. C* **71**, 1777 (2011).

Myung, Y.S., *Phys. Rev. D* **84** (2), 024048 (2011).

Myung, Y.S., Moon, T., and Son, E.J., *Phys. Rev. D* **83** (12), 124009 (2011).

Näf, J. and Jetzer, P., *Phys. Rev. D* **84** (2), 024027 (2011).

Nashed, G.G.L., *Eur. Phys. J. Plus* **133**, 18 (2018).

Nashed, G.G.L. and Capozziello, S., *Phys. Rev. D* **99**, 104018 (2019).

Nishizawa, A., *Phys. Rev. D* **98** (10), 104037 (2018).

Nojiri, S. and Odintsov, S.D., *Phys. Lett. B* **779**, 425 (2018).

Novikov, I.D. and Thorne, K.S., in *Black holes (Les astres occlus)*, Eds. C. DeWitt and B. DeWitt (Gordon and Breach, N.Y., 1973).

Olmo, G.J. and Rubiera-Garcia, D., *Phys. Rev. D* **84** (12), 124059 (2011).

Olmo, G.J. and Rubiera-Garcia, D., *Phys. Rev. D* **86** (4), 044014 (2012).

Olmo, G.J. and Rubiera-Garcia, D., *Eur. Phys. J. C* **72**, 2098 (2012).

Olmo, G.J., Rubiera-Garcia, D., and Sanchez-Puente, A., *J. Phys. Conf. Ser.* 012042 (2015).

Olmo, G.J., Rubiera-Garcia, D., and Sanchez-Puente, A., *Universe* **1**, 173 (2015).

Olmo, G.J., Rubiera-Garcia, D., and Sanchez-Puente, A., *Eur. Phys. J. C*, **76**, 143 (2016).

Oppenheimer, J.R. and Volkoff, G.M., *Phys. Rev.* **55**, 374 (1939).

Orellana, M. *et al.*, *Gen. Relativ. Gravit.* **45**, 771 (2013).

Orosz, J.A. *et al.*, *Astrophys. J.* **742**, 84 (2011).

Özel, F. *et al.*, *Astrophys. J.* **757**, 55 (2012).

Page, D.N. and Thorne, K.S., *Astrophys. J.* **191**, 499 (1974).

Pandharipande, V.R. and Ravenhall, D.G., *NATO Advanced Science Institutes (ASI) Series B*, **205**,103 (1989).

Pani, P. *et al.*, *Phys. Rev. D* **84** (10), 104035 (2011).

Pardo, K. *et al.*, *J. Cosm. Astropart. Phys.* **07**, 048 (2018).

Paul, N., Nil Chakrabarty, S., and Bhattacharya, K., *J. Cosm. Astropart. Phys.* **10**, 009 (2014).

Pearson, J.M., Goriely, S., and Chamel, N., *Phys. Rev. C.* **83** (6), 065810 (2011).

Pearson, J.M., Chamel, N., Goriely, S., and Ducoin, C., *Phys. Rev. C.* **85** (6), 065803 (2012).

Penrose, R., *Phys. Rev. Lett.* **14**, 57 (1965).

Pérez, D., Romero, G.E., and Perez Bergliaffa, S.E., *Astron. Astrophys.* **551**, A4 (2013).

Pérez, D., Romero, G.E., and Perez Bergliaffa, S.E., *IJTP* **53**, 734 (2014).

Pérez, D., Armengol, F.G.L., and Romero, G.E., *Phys. Rev. D* **95**, 104047 (2017).

Psaltis, D. *et al.*, *Phys. Rev. Lett.* **100** (9), 091101 (2008).

Rezzolla, L., Zanotti, O., and Font, J.A., *Astron. Astrophys.* **412** (3), 603 (2003).

Robinson, D.C., *Phys. Rev. Lett.* **34**, 905 (1975).

Rodrigues, M.E. *et al.*, *Phys. Rev. D* **94** (2), 024062 (2016).

Romero, G.E., arXiv:1210.2427R (2013).

Ruiz, M., Shapiro, S.L., and Tsokaros, A., *Phys. Rev. D.* **97**, 021501 (2018).

Safonova, M., Torres, D.F., and Romero, G.E., *Phys. Rev. D* **65** (2), 023001 (2002).

Sandage, A. *et al.*, *Astrophys. J.* **146**, 316 (1966).

Schwarzschild, K., *Sitzungsberichte der Königlich Preußischen Akademie der Wissenschaften (Berlin)*, 189 (1916).

Seymour, B. and Yagi, K., *Phys. Rev. D* **98**, 124007 (2018).

Shakura, N.I., *Astronomicheskii Zhurnal* **49**, 921 (1972).

Shakura, N.I. and Sunyaev, R.A., *Astron. Astrophys.* **24**, 337 (1973).

Shapiro, I.I., *Phys. Rev. Lett.* **13**, 789 (1964).

Sheykhi, A., *Phys. Rev. D* **86** (2), 024013 (2012).

Shklovsky, I.S., *Astrophys. J. Lett.* **148**, L1 (1967).

Silbar, R.R. and Reddy, S., *Am. J. Phys.* **72**, 892 (2004).

Slaný, P. and Pokorná, M. and Stuchlík, Z., *Gen. Relat. Gravit.* **45**, 2611 (2013).

Sotiriou, T.P. and Faraoni, V., *Rev. Mod. Phys.* **82**, 451 (2010).

Spruit, H.C. *Lecture Notes in Physics* (Springer Verlag, Berlin, 2010).

Starobinsky, A.A., *JETPL* **86**, 157 (2007).

Staykov, V.K. *et al.*, *J. Cosm. Astropart. Phys.* **10**, 006 (2014).

Stuchlík, Z. and Hledík, S., *Phys. Rev. D* **60** (4), 044006 (1999).

Stuchlík, Z. and Slaný, P., *Phys. Rev. D* **69** (6), 064001 (2004).

Sultana, J. and Kazanas, D., *Gen. Relativ. Grav.* **50**, 137 (2018).

Teppa Pannia, F.A. *et al.*, *Gen. Relativ. Gravit.* **49**, 25 (2017).

Tolman, R.C., *Phys. Rev.* **55**, 364 (1939).

Vainio, J. and Vilja, I., *Gen. Relativ. Gravit.* **49**, 99 (2018).

Webster, B.L and Murdin, P., *Nature* **235**, 37 (1972).

Wei, S.-W. and Liu, Y.-X., *Phys. Rev. D* **98** (2), 024042 (2018).

Will, C.M., *Liv. Rev. Relativ.* **17**, 4 (2014).

Wojnar, A., *Eur. Phys. J. C* **78**, 421 (2018).

Yagi, K. *et al.*, *Phys. Rev. D* **87** (8), 084058 (2013).

Yazadjiev, S.S. *et al.*, *J. Cosm. Astropart. Phys.* **6**, 003 (2014).

Yazadjiev, S.S. *et al.*, *Phys. Rev. D* **91** (8), 084018 (2015).

Zhang, F. and Saha, P., *Astrophys. J.* **849**, 33 (2017).

Zhang, F. *et al.*, *Astrophys. J.* **874**, 121 (2019).

Chapter 3

The Pseudo-Complex General Relativity: Theory and Observational Predictions

Peter O. Hess[*] and Thomas Boller[†]

Instituto de Ciencias Nucleares
Universidad Nacional Autónoma de Mexico (UNAM)
Circuito Exterior, C.U., A.P. 70-543
04510 Ciudad de Mexico, Mexico
hess@nucleares.unam.mx
[†]*Max-Planck Institute for Extraterrestial Physics (MPIEP)*
Giessenbachstrasse, D-85748 Garching, Germany
Frankfurt Institute for Advanced Studies (FIAS)
Johann Wolfgang Goethe Universität (JWGU)
Ruth-Moufang-Str. 1, 60438 Frankfurt am Main, Germany
bol@mpe.mpg.de

A review is presented on the *pseudo-complex General Relativity* (pcGR) and its observational consequences. Significant differences to the *General Relativity* (GR) appear only in the regime of strong gravitational fields. Robust differences between GR and pcGR are predicted in the appearance of an accretion disk. Finally, some words are spent on the observation of gravitational waves.

3.1. Introduction

The theory of *General Relativity* (GR) is well established and has passed a series of observational tests [Will (2006)], mainly within the solar system. There are others, for example the Hulse–Taylor pulsar [Hulse and Taylor (1975)] and the recent observation of gravitational waves [Abbott *et al.* (2016)]. Only the last one is connected with strong gravitational fields.

67

Despite the success of GR, it is not clear at all if it is complete. In extreme situations, such as in a strong gravitational field, other effects may be added, for example quantum fluctuations. Also a minimal length may be added (see for example [Caianiello (1981)]), simulating a granulation of space.

Another possibility to generalize GR is to apply an *algebraic extension* of the coordinates to a different type, as for example to complex ones [Mantz (2008)]. In [Hess and Greiner (2009); Caspar *et al.* (2012); Hess, Schäfer, and Greiner (2015); Hess and Greiner (2017)], a new theory was proposed using a *pseudo-complex* (pc) algebraic extension of GR. This theory will be exposed in the next section and shown that it requires the presence of a dark energy, which accumulates at smaller distances to a large mass. This leads to the conjecture that *the mass not only curves spacetime but it also changes vacuum properties*. When the parameters of the theory are chosen appropriately, this will enable us to avoid the *event horizon*!

Why do we want to avoid the event horizon? One reason is that the quantum effects will lead to a dark energy with repulsive behavior. If it is strong enough it could halt any collapse to a singularity. Second, the event horizon due to the large mass in GR is a very special one: It excludes the access of a *nearby* observer to a part of the space and also it produces the coordinate singularity called *the event horizon*, a very unsatisfactory consequence. It indicates the possibility that GR is not complete. This gave us the motivation to investigate possible extensions of GR, of which an algebraic one is a possibility.

In this contribution we will explain the theory in Sec. 3.2. In Sec. 3.3 some observational predictions will be presented. As an example, accretion disks will be simulated and a distinct feature will be presented: The appearance of a dark ring followed with a bright inner one. This feature is independent of the particular disk model used, while the distances to the center might change due to a different fall off of the dark energy as a function of the radial distance. Finally, some words on gravitational waves will be given and a discussion of the observed events will be presented.

3.2. The Pseudo-Complex General Relativity

There has been several former attempts to extend the *pseudo-complex General Relativity* (pcGR). One of those was proposed by A. Einstein [Einstein (1945, 1948)], who introduced a complex metric, where the real part is associated with the standard one and the imaginary part to the Maxwell tensor. M. Born [Born (1938, 1949)] was, to our knowledge, the first scientist to propose an algebraic extension, namely using complex coordinates where the imaginary part is proportional to the linear momentum of a particle. His motivation was to unify Quantum Mechanics with the GR, elevating the coordinates and momenta in GR to the same level. The main problem is the dependence of the coordinates on the mass of a particle. In the same line of thoughts, E. R. Caianiello proposed [Caianiello (1981)] a new length element, consisting in two parts: The first term has the same structure as in GR and the additional, second element, depends on the square of infinitesimal four-velocities. Due to dimensional reasons, a minimal length squared had to be introduced in order to keep the dimensions correct. This was related to a maximal acceleration. In [Mantz (2008)] the use of a complex algebraic extension was reconsidered and extended.

In [Kelly and Mann (1986)] all possible algebraic extensions were investigated in the limit of weak gravitational fields. As a main result, the authors proved that only one algebraic extension does not lead to ghost and/or tachyon solutions, namely the pseudo-complex one (named differently in [Kelly and Mann (1986)]), which is the reason to consider only these coordinates.

In [Hess and Greiner (2009)] the GR was algebraic extended to pc-coordinates, with the structure

$$X^\mu = x^\mu + Iy^\mu, \quad I^2 = 1. \tag{3.1}$$

Alternatively one can define

$$\sigma_\pm = \frac{1}{2}(1 \pm I) \rightarrow X^\mu = X_+^\mu \sigma_+ + X_-^\mu \sigma_-$$

with

$$\sigma_\pm^2 = 1, \quad \sigma_+\sigma_- = 0. \tag{3.2}$$

Especially the last property shows that two non-zero elements multiplied can lead to a zero, thus the variables do not form a field but rather a ring. Because the $\sigma_\pm$ behave as projectors, one can apply mathematical manipulations independently in each component, which are also called the components of the *zero divisor basis*. Details can be retrieved in Hess, Schäfer, and Greiner (2015).

In extending GR to pcGR, a pc-metric is introduced

$$g_{\mu\nu} = g_{\mu\nu}^+ \sigma_+ + g_{\mu\nu}^- \sigma_-, \tag{3.3}$$

where we restrict to symmetric ones (i.e., no torsion).

The new length element squared acquires the form

$$
\begin{aligned}
d\omega^2 = g_{\mu\nu}dX^\mu dX^\mu &= g_{\mu\nu}^+ dX_+^\mu dX_+^\mu \sigma_+ + g_{\mu\nu}^- dX_-^\mu dX_-^\mu \sigma_- \\
&= g_{\mu\nu}^s (dx^\mu dx^\nu + dy^\mu dy^\nu) + 2g_{\mu\nu}^a dx^\mu dy^\nu \\
&\quad + I[g_{\mu\nu}^a(dx^\mu dx^\nu + dy^\mu dy^\nu) + 2g_{\mu\nu}^s dx^\nu dy^\nu],
\end{aligned} \tag{3.4}
$$

where $g_{\mu\nu}^s = \frac{1}{2}(g_{\mu\nu}^+ + g_{\mu\nu}^-)$ and $g_{\mu\nu}^a = \frac{1}{2}(g_{\mu\nu}^+ - g_{\mu\nu}^-)$. For the case of $g_{\mu\nu}^- = g_{\mu\nu}^+$ and $dy^\mu = lu^\mu$, one recovers the length element proposed by E. R. Caianiello [Caianiello (1981)] .

The two zero-divisor components are connected via the condition of a real infinitesimal length element squared, i.e., that a particle only moves along paths with a real length. This demands that the pseudo-imaginary component of (3.4) is zero (we use $I = (\sigma_+ - \sigma_-)$):

$$(\sigma_+ - \sigma_-)(g_{\mu\nu}^+ dX_+^\mu dX_+^\nu - g_{\mu\nu}^- dX_-^\mu dX_-^\nu) = 0. \tag{3.5}$$

Only if the background metric is Minkowskian, i.e. $g_{\mu\nu}^\pm = \eta_{\mu\nu}$, the constraint (3.5) leads to the identification of y^μ to lu^μ and only in this case the result coincides with [Caianiello (1981)]. In order to get a solution for y^μ the constraint equation (3.5) has to be resolved explicitly, which is a difficult task.

The action within pcGR is defined as

$$S = \int dx^4 \sqrt{-g}(\mathcal{R} + 2\alpha), \tag{3.6}$$

where α may depend on the radial distance when a central problem is considered, or even on the azimuthal angle when the central object is rotating.

Using the constraint (3.5) leads to the equations of motion

$$\mathcal{R}_{\mu\nu}^{\pm} - \frac{1}{2}g_{\mu\nu}^{\pm}\mathcal{R}_{\pm} = \lambda(\dot{X}_{\mu}^{\pm}\dot{X}_{\nu}^{\pm}) + \alpha g_{\mu\nu}^{\pm}$$

$$= \lambda u_{\mu}u_{\nu} + \lambda(\dot{y}_{\mu}\dot{y}_{\nu} \pm u_{\mu}\dot{y}_{\nu} \pm u_{\nu}\dot{y}_{\mu}) + \alpha g_{\mu\nu}^{\pm}$$

$$= 8\pi T_{\pm\ \mu\nu}^{\Lambda}, \tag{3.7}$$

where the derivative of the coordinate $(X_{\pm}^{\mu})$ with respect to $s = ct = t$ was expressed in terms of the pseudo-real and pseudo-imaginary component of the X^{μ}. The right hand side of the Einstein equations is identified with an energy–momentum tensor. For a complete solution, the constraint has to be resolved which turns out to be quite difficult.

Up to here we applied an abstract, theoretical point of view. The question is now: What is the origin of the energy–momentum tensor? Hints are obtained considering the results of semi-classical Quantum Mechanics [Visser (1996)]: Assuming a fixed background metric, as the Schwarzschild one, and determining the vacuum fluctuations provoked by the non-flat metric.

In [Visser (1996)] the dark energy behaves as

$$\rho_{\Lambda}^{V} = -3p_{\infty}\left(\frac{r_s}{r}\right)^6 \frac{\left[40 - 72\left(\frac{r_s}{r}\right) + 33\left(\frac{r_s}{r}\right)^2\right]}{\left(1 - \frac{r_s}{r}\right)^2}, \tag{3.8}$$

where p_{∞} is defined as $\hbar/(90(16\pi)^2(2m)^4)$ and the upper index "V" refers to "Visser", the author of the publication.

As noted, the energy density increases with lower radial distances and explode at the event horizon (because the background metric is fixed, there is still a horizon). For large radial distances, the dark energy density behaves as $1/r^6$. Because the gravitational field is very strong near the event horizon, the applicability of the semi-classical approach is questionable. Nevertheless, the useful information is the increase of the dark energy toward the center. We take this result as a justification to assume a dark energy density which decreases with increasing r, such that the effects are not yet seen in solar system experiments but still strong enough to hope for some observable consequences. Because of the assumptions made, the pcGR acquires from here on a phenomenological feature. We assume an *ad hoc* fall off of the density proportional to $1/r^5$, which is finite at the place of the so-called event horizon.

In a more general ansatz, the dark energy density may fall off as

$$\rho_\Lambda = \frac{B}{8\pi r^n}, \quad n \geq 5, \tag{3.9}$$

where the index Λ refers to the dark energy. This affects the structure of the Kerr metric, which for $n = 5$ is given by

$$g^{\mathrm{K}}_{00} = \frac{r^2 - 2mr + a^2 \cos^2 \vartheta + \frac{B}{2r}}{r^2 + a^2 \cos^2 \vartheta},$$

$$g^{\mathrm{K}}_{11} = -\frac{r^2 + a^2 \cos^2 \vartheta}{r^2 - 2mr + a^2 + \frac{B}{2r}},$$

$$g^{\mathrm{K}}_{22} = -r^2 - a^2 \cos^2 \vartheta,$$

$$g^{\mathrm{K}}_{33} = -(r^2 + a^2) \sin^2 \vartheta - \frac{a^2 \sin^4 \vartheta \left(2mr - \frac{B}{2r}\right)}{r^2 + a^2 \cos^2 \vartheta},$$

$$g^{\mathrm{K}}_{03} = \frac{-a \sin^2 \vartheta \, 2mr + a\frac{B}{2r} \sin^2 \vartheta}{r^2 + a^2 \cos^2 \vartheta}, \tag{3.10}$$

where the index K refers to Kerr. For $B = 0$ the Kerr solution of GR is obtained.

The minimum of g_{00} is calculated for an arbitrary n-dependence and given by

$$r_0 = \left[\frac{(n-2)}{2(n-3)(n-4)} \frac{B}{m}\right]^{\frac{1}{(n-3)}}. \tag{3.11}$$

Requiring that there is no event horizon implies that $g_{00} > 0$, which leads to the lower limit of B (see Eq. (3.13) below). For $n = 5$ the r_0 is at $(2/3)$ of the Schwarzschild radius, while for $n = 6$ it shifts to the larger value of $(3/4)$ of the Schwarzschild radius.

For $n = 5$ one obtains $B > \frac{64}{27}m^3$, though for practical purposes we just use $B = \frac{64}{27}m^3$ which gives a zero for the minimal value of g_{00} at the mentioned two thirds of the Schwarzschild radius. The rotational parameter a is in units of m and ranges as usual from

$0m$ to $1m$. When n is different from 5, the $B/(2r)$ contribution in (3.10) changes to

$$\frac{B}{(n-3)(n-4)r^{n-4}}. \tag{3.12}$$

The requirement of no event horizon results in a minimal value for B, i.e.,

$$B > B_{min} = \left[\frac{2(n-3)(n-4)}{(n-2)}\right] \left[\frac{2(n-3)}{(n-2)}\right]^{(n-3)} m^{(n-2)}. \tag{3.13}$$

For $n = 5$ this gives $B_{min} = \frac{64}{27}m^3$, while for $n = 6$ it is $\frac{81}{8}m^4$, i.e., the value is significantly larger than for $n = 5$.

The B measures the coupling of the mass to the dark energy. The larger the mass the larger the coupling. The factor b in $B = bm^{(n-2)}$ can be considered as a universal parameter, once n is known.

In order to understand the differences between GR and pcGR seen for quasi-periodic objects (QPOs), it is necessary to discuss the effects on the redshift. In Fig. 3.5 the left panel is the redshift for an azimuthal angle of 90°. As shown, the curve of pcGR is nearly a copy of the one in GR, shifted to lower values in r. While in GR the curve tends to infinity the one in pcGR tends to *very large* values, i.e. the central object behaves like a real black hole. The situation is different at the poles, where the redshift in pcGR is not larger than 1, i.e., there is a hope that light emission from the poles can be measured, if it is not hidden by a jet.

3.3. Observational Consequences

In order to understand the observational consequences, presented in this section, some of the properties of a particle in a circular orbit are recalled.

In Fig. 3.1 the pcGR orbital frequency of a particle in a circular orbit is depicted as a function in r, for a near maximal rotational parameter a. The orbital frequency in pcGR is always below the one of GR. The reason is the effectively decreasing gravitational constant due to the accumulation of dark energy at lower radial distances.

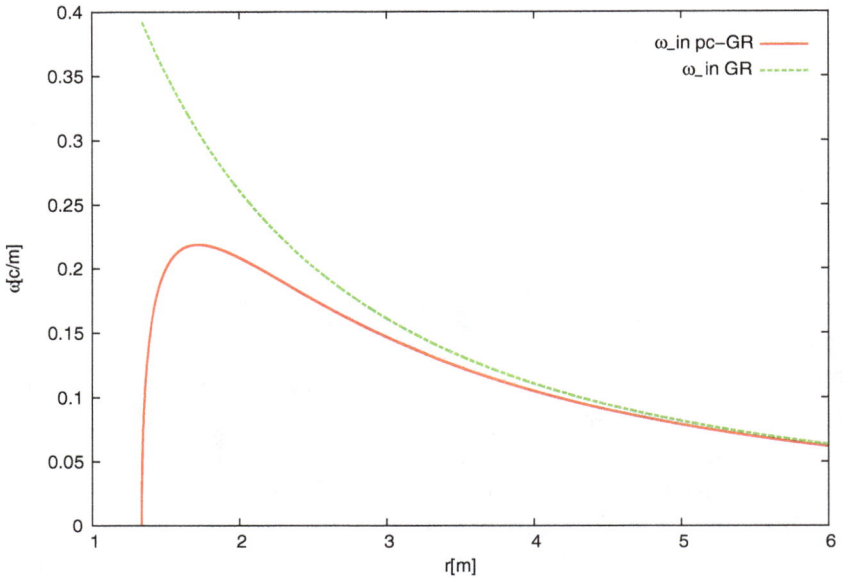

Fig. 3.1   Orbital frequency, in hertz, is plotted as a function in r, for a particle in a circular orbit. The rotational parameter is $a = 0.995m$.

In pcGR the curve shows a maximum after which it falls off and approaches zero at the surface of the star (where $g_{00} = 0$). The maximum is at approximately $1.72m$ ($n = 5$). Near the maximum two neighboring orbitals have a very similar orbital frequency and thus friction is substantially lowered. When an accretion disk is considered, this is the position where the disk is less excited and the intensity of the light emission acquires a minimum. When $n > 5$, the position of the minimum changes to

$$r_{\omega_{\max}} = \left[\frac{n(n-2)}{6(n-3)(n-4)} \frac{B}{m} \right]^{\frac{1}{n-3}}. \tag{3.14}$$

For $n > 5$ the dark minimum shifts to larger values of r at $\omega_{\max}$. While for $n = 5$, the position is at $1.72m$, it is $1.89m$ for $n = 6$, slightly larger.

The important result is that a change in n changes the position but not the characteristic of the appearance of a minimum, thus, this

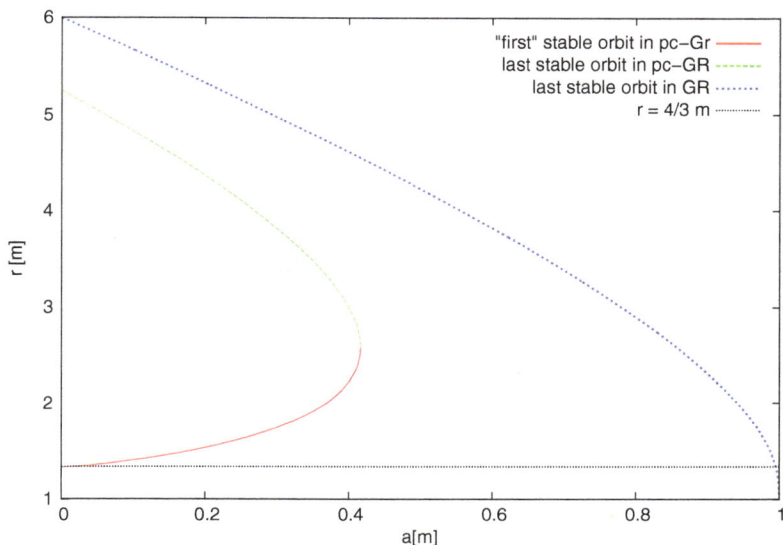

Fig. 3.2 The position of the *last stable orbit* as a function of the rotational parameter *a*.

structure is robust and from a known position of the minimum the value of n can be deduced.

In what follows, we will restrict to $n = 5$, but from the above discussion it should be clear how the values change once n is different.

In Fig. 3.2 the last stable orbit is plotted versus the rotational parameter a. At $a = 0$ (Schwarzschild) the last stable orbit is at the known distance of $r = 6m$, while for pcGR it is a little bit further in. This permits to release more gravitational energy, which is distributed within the disk, and thus the disk must be brighter. The situation is similar when a is increased, until it is a bit larger than 0.416. For larger a stable orbits exist within pcGR until the surface of the star and they pass over the position of the maximum in Fig. 3.1 and, therefore, a minimum in the emission profile appears.

3.3.1. *Simulations of accretion disks*

As discussed above, when circular particle orbits were discussed, for the case of a slowly rotation black hole, GR and pcGR are quite similar, except that pcGR will show a larger light emission intensity of the disk.

For rapidly rotating black holes, however, the situation changes drastically. Stable orbits exist until the surface of the star and pass over the maximum of the orbital frequency in r. At the radial distance of the maximum, the emission of light is low and the disk should show a dark ring, which is followed at smaller distances by a bright one, a consequence of the rapid change of the orbital frequency.

Note that this discussion is of qualitative nature and so does not depend much on how strong the dark energy density falls off in r. Changing the power of n only changes the positions of the dark and the bright ring, as was shown above.

For the simulation of the accretion disk it is not of importance to have a very realistic model, because all models should show the structure discussed above. For simplicity, we use a quite simple model, proposed in 1974 by D. N. Page and K. S. Thorne [Page and Thorne (1974)]. The main assumptions are:

 i) The disk is thin, optically opaque and infinitely extended.
ii) The emission of light is only through photons.

The theory uses conservation laws for mass, energy and angular momentum. With these few assumptions and the use of conservation laws, the disk can be simulated. All unknown properties, e.g. the viscosity of the disk, the accretion rate and the rotational parameter a are hidden within the parameters. Another advantage emerges from [Vincent *et al.* (2011); Gyoto (2015)], where a numerical routine was published, which uses the method of *raytracing* and provides the model of D. N. Page and K. S. Thorne. This routine allows to change the metric.

The disk model and the raytracing method are independent of the metric used. In the raytracing method, rays are *followed back* from the detector to the accretion disk, using the Hamilton–Jacobi formalism. It leads to a set of differential equations, which have to be solved numerically. The metric appears as a general function.

In Fig. 3.3 the simulation for $a = 0.9$ is shown, for an inclination angle of observer-disk of 80°. The left figure depicts the result within

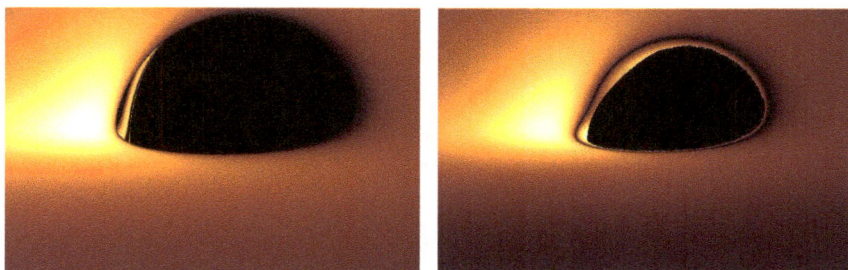

Fig. 3.3 Infinite, counter-clockwise rotating geometrically thin accretion disk around static and rotating compact objects viewed from an inclination of $80°$. The left panel shows the original disk model as obtained by [Page and Thorne (1974)]. The right panel shows the results within pcGR. The spin parameter is $a = 0.9$. a is given in units of m.

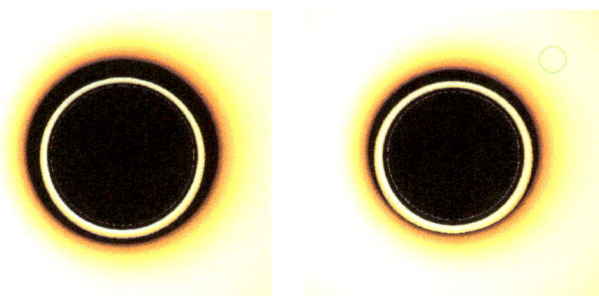

Fig. 3.4 Infinite, counter-clockwise rotating geometrically thin accretion disk around static and rotating compact objects viewed from an inclination of $10°$. The left panel shows the original disk model as obtained by [Page and Thorne (1974)]. The right panel shows the results within pcGR. The spin parameter is $a = 0.0$, i.e., the Schwarzschild case.

GR and the right figure the one of pcGR. The difference is clearly seen: In pcGR the disk is brighter and a dark fringe is followed by a bright ring further in. As can be seen, the GR does not show this structure. The GR case does show, however, a remnant from the first Einstein ring, where the light ray completes a full turn around the black hole before heading to the detector. Consistently the disk appears brighter in pcGR than in GR.

In Fig. 3.4 the same is shown, but now for the Schwarzschild case ($a = 0$) and an inclination angle of $10°$, i.e., nearly from above.

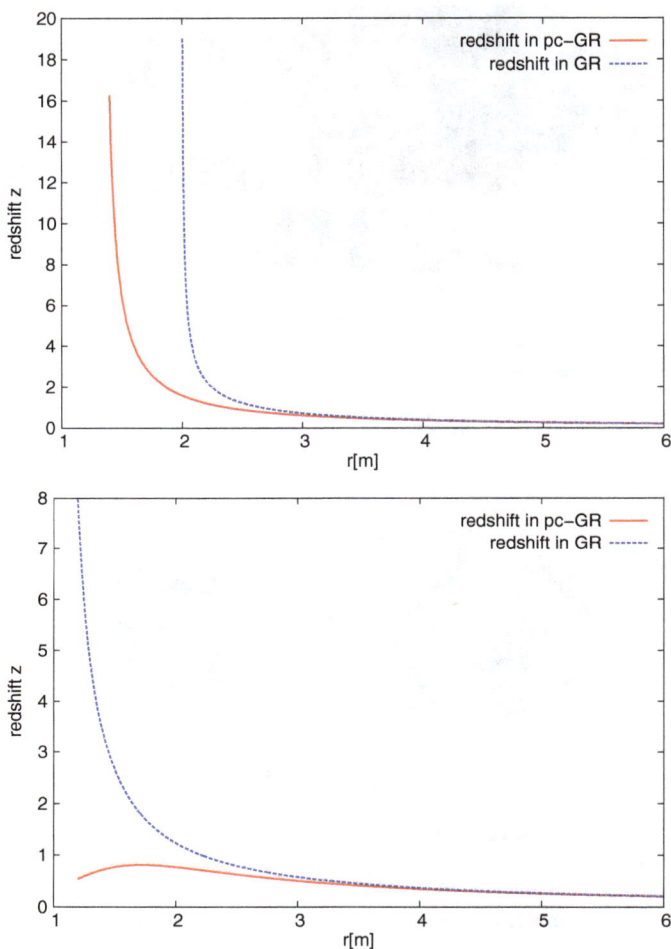

Fig. 3.5   The upper panel depicts the redshift at the azimuthal angle of 90°. The inner curve is pcGR while the outer one is GR. The lower panel depicts the redshift at zero azimuthal angle (at the poles). The lower curve is pcGR while the upper one is GR.

The left panel is GR and the right one is pcGR. The structure is now very similar, because stable orbits are not present at the maximum of the orbital frequency in pcGR. Because in pcGR the stable orbits reach further in, the disk appears brighter.

Unfortunately, there is no evidence of an accretion disk in SgrA*, the central black hole in our galaxy. Actually measurements have been completed by the collaboration of the Event Horizon Telescope [ETH (2016)], which were published recently.

In case no accretion disk is found, one can nevertheless use the results for an in-falling cloud, which should follow the structure of the disk result, namely first lighting up, then lowering its emission up to where the minimum in the accretion disk would be, followed by a last brightening before completely disappearing.

3.4. Gravitational Waves

The first observation of gravitational waves have been reported in [Abbott *et al.* (2016)]. The source was interpreted as the merger of two black holes of about 30 solar masses each, at a distance of approximately 1.3 billion light years. This measurement led to the Nobel Prize for three physicists in 2017.

In [Hess (2016)] a very simple model was applied to study the 2016 event. The two stars in the pre-merger phase where approximated by two mass points, which is not a very good assumption because the gravitational constant varies strongly with the stars, assuming to orbit each other in close proximity.

The dependence of frequency on the radial distance is given by

$$R^2 = \left(\frac{G\widetilde{M}}{\omega_s^2}\right)^{\frac{2}{3}} [F_\omega(r)]^{\frac{2}{3}}. \tag{3.15}$$

For the chirping mass the following relation was found:

$$\mathcal{M}_c = \widetilde{M}_c F_\omega(r) = \frac{c^3}{G}\left[\frac{5}{96\pi^{\frac{8}{3}}}\frac{df_{\mathrm{gw}}}{dt}f_{\mathrm{gw}}^{-\frac{11}{3}}\right]^{\frac{3}{5}}, \tag{3.16}$$

where $F_\omega(r)$ is a function with a maximum at $1.72m$ (for $n = 5$) and it gets zero at $r = \frac{4}{3}m$. Using the observed frequency and the change of it, near the touching configuration, this function becomes very small, demanding higher masses. (In [Hess (2016)] the formula (3.16) contained errors, which are eliminated now. The errors had no consequence for the final results.)

This model suggests that the event rather corresponds to the merger of two black holes of the size of several thousand solar masses and at a distance corresponding to the early time of the universe. In this picture, the event may be related to the merger of two black holes, whose former galaxies collided.

In [Nielsen and Birnholz (2018, 2019)] this conjecture was reanalyzed, using an n-dependence of the dark energy density. The premerger phase was considered when both stars are still far apart such that Newton's theory can be applied and GR, pcGR agree. The two stars were approximated by two mass points. In addition the ringdown frequency was calculated, considering a test particle moving in a light ring. The authors claim that pcGR with the $n = 5$ dependence of the dark energy seems to be excluded, i.e., n is probably larger than 5. Also, the mass should be similar to the one deduced in [Abbott *et al.* (2016)].

In order to understand a possible glitch in their analysis, one has to keep in mind Fig. 3.1, where the orbital frequency is plotted versus the distance. For a fixed frequency (the two horizontal dashed lines) *there are two intersections*, one at small and the other at large distances. For large distances the two theories are very similar, i.e., when one starts with two 30 solar mass objects, which fits the GR prediction and both theories give similar results and one has to increase n in pcGR, in order to get near the observation. I.e., when one starts with two stars of 30 solar masses, the results of [Nielsen and Birnholz (2018, 2019)] would be obtained.

But what happens when the stars are much heavier? At large distances the frequency squared scales with the mass and a heavier mass implied a higher frequency, too large for observation. The horizontal dashed lines in Fig. 3.1 would intersect the curve at $r \gg m$, long before merger. Only when the two stars are nearly touching, the observed frequency is obtained in the left-lower part of the curve in Fig. 3.1. However, because the effective gravitational constant varies significantly within a star, the two-point approximation is not valid anymore and the wave equations have to be solved numerically [Chandrasekhar (1983)].

In conclusion: The observation of gravitational waves permits two interpretations: One with 30 solar masses which fits well with GR but requires to increase n in pcGR, or much larger masses, which only near the touching point reaches the observed frequency.

We are currently investigating the stability of a Schwarzschild pcGR metric, using the method explained in [Chandrasekhar (1983)], in order to test the results of [Nielsen and Birnholz (2018, 2019)]. This leads to the Zerrilli equations [Zerrilli (1970)], from which the ring-down frequencies can be determined. This method does not depend on the assumption of the existence of a light ring. This explicit calculation is the only way to find out if in pcGR the wave form can be reproduced, using the second scenario mentioned above.

3.5. Conclusions

The most recent status of the *pseudo-complex General Relativity* was presented. A reasoning was given why pseudo-complex coordinates are the only possible algebraic extension of the standard theory. After an explanation of pcGR the rest of this contribution concentrated on the observational consequences such as:

a) quasi-periodic oscillations in accretion disks of AGN's and stellar black holes,
b) structure of an accretion disk and
c) gravitational waves.

In a) there are some hints that GR has problems, though alternative explanations are possible. The differences between GR and pcGR, however, are more robust and provide the only observational difference to be seen in near future. The observation also provides, in case the dark fringe followed by a bright ring is observed, the determination of the exponent n for the fall-off of the dark energy. In case of the gravitational waves, the discussion is ongoing whether pcGR provides the correct description of the observation. Hints are found that $n = 5$ is not large enough, but the conclusions are made on the assumptions on the validity of a two-point approximation and that there exists a range at large relative distances where the two theories are the same and the particular frequency observed is generated. We showed that when the masses are very large, in the range of great distances the frequency generated is too high and only near the merger it reaches the observed range. Only a numerical calculation can shed more light onto this problem.

Note added in proof

Since this contribution was written, a lot has happened, such as the first observation of the black hole shadow in M87 by the EHT collaboration. These new developments are addressed in [Hess and López-Moreno (2019)]. We refer to this publication for more information.

Acknowledgment

P.O.H. acknowledges support from DGAPA-PAPIIT (IN100418).

References

Abbott B.P. *et al.*, (*LIGO Scientific Collaboration and Virgo Collaboration*), *Phys. Rev. Lett.* **116**, 061102 (2016).
Born, M., *Proc. Roy. Soc. A* **165**, 291 (1938).
Born, M., *Rev. Mod. Phys.* **21**, 463 (1949).
Caianiello, E.R., *Nuovo Cim. Lett.* **32**, 65 (1981).
Caspar, G., Schönenbach, T., Hess, P. O., Schäfer, M., and Greiner, W., *Int. J. Mod. Phys. E* **21**, 1250015 (2012).
Chandrasekhar, S., *The Mathematical Theory of Black Holes*, International Series of Monographs in Physics, Vol. 69 (Oxford University Press, Oxford, 1983).
Einstein, A., *Ann. Math. Second Ser.* **46**, 578 (1945).
Einstein, A., *Rev. Mod. Phys.* **20**, 35 (1948).
Event Horizon Telescope, www.eventhorizontelescope.org (2016).
Gyoto Manual, www.gyoto.obspm.fr/GyotoManual.pdf (2015).
Hess, P.O. and Greiner, W., *Int. J. Mod. Phys. E* **18**, 51 (2009).
Hess, P.O. and López-Moreno, E., *Universe* **5**, 191 (2019).
Hess, P.O., Schäfer, M., and Greiner, W., *Pseudo-Complex General Relativity* (Springer, Heidelberg, 2015).
Hess, P.O., *Mon. Not. Roy. Astron. Soc.* **462**, 3026 (2016).

Hess, P.O. and Greiner, W., Pseudo-complex general relativity: Theory and observational consequences, in *Centennial of General Relativity: A Celebration*, Ed. C.A. Zen Vasconcellos (World Scientific, Singapore, 2017).

Hulse, R.A. and Taylor, J.H., *Astrophys. J.* **195**, L51 (1975).

Kelly, P.F. and Mann, R.B., *Class. Quantum Gravity* **3**, 705 (1986).

Mantz, C.L.M. and Prokopec, T., arXiv:0804.0213 (2008); Mantz, C.L.M. and Prokopec, T., *Found. Phys.* **41**, 1597 (2011).

Nielsen, A.B. and Birnholz, O., *Astron. Nachr.* **339**, 298–305 (2018).

Nielsen, A.B. and Birnholz, O., *Astron. Nachr.* **340**, 116–120 (2019).

Page, D.N. and Thorne, K.S., *Astrophys. J.* **191**, 499 (1974).

Vincent, F.H., Paumard, T., Gourgoulhon, E., and Perrin, G., *Class. Quantum Grav.* **28**, 225011 (2011).

Visser, M., *Phys. Rev. D* **54**, 5116 (1996).

Will, C.M., *Living Rev. Relativ.* **9**, 3 (2006).

Zerrilli, F.J., *Phys. Rev. D* **2**, 2141 (1970).

Chapter 4

Dense Baryonic Matter in the Cores of Neutron Stars

William M. Spinella* and Fridolin Weber[†]

*Department of Sciences, Wentworth Institute of Technology
550 Huntington Avenue, Boston, MA 02115, USA
spinellaw@wit.edu
[†]Department of Physics, San Diego State University
5500 Campanile Drive, San Diego, California 92182, USA
Center for Astrophysics and Space Sciences
University of California at San Diego
La Jolla, California 92093, USA
fweber@sdsu.edu, fweber@ucsd.edu

We introduce the relativistic mean-field approximation for determining the equation of state of hadronic matter along with a number of associated nuclear parameterizations. Constraints on the NS equation of state from nuclear physics and NS observations are discussed and used to eliminate several popular nuclear parameterizations. Finally we determine the hadronic equation of state with the inclusion of hyperons and systematically investigate the parameter space of the meson–hyperon coupling constants.

4.1. Introduction

Neutron stars are among the densest astrophysical objects in the universe. These objects were first proposed as a possible endpoint of stellar evolution by Baade and Zwicky in 1933 shortly after the discovery of the neutron by Chadwick [Baade and Zwicky (1934); Chadwick (1932)]. Neutron stars are formed in core-collapse super-

novae. These occur in stars that are massive enough to fuse iron
(^{56}Fe) in their cores, with masses typically between 10–25 times that
of our Sun ($M_\odot$). Nuclear fusion in the core of a star is an exothermic process, and the energy released provides the pressure necessary
to balance the gravitational attraction of the star's mass. This balance is known as hydrostatic equilibrium. However, nuclear fusion in
the stellar core ceases with ^{56}Fe because fusing heavier elements is
an endothermic rather than exothermic process. Beyond this point
in its evolution the stellar core must be supported against gravity
by electron degeneracy pressure alone, which gravity overcomes at
the Chandrasekhar limit (about $1.4\,M_\odot$). If continued nuclear fusion
from shell burning deposits enough mass on the core that it exceeds
this limit the core will collapse until halted by neutron degeneracy
pressure. A subsequent rebound of the remaining core and neutrino
production due to electron capture fuel an outward explosion of matter known as a supernova, leaving behind a neutron star. If the core
of a star does not become massive enough to fuse sufficient ^{56}Fe and
overcome electron degeneracy the star will end its evolution as a less
dense compact object known as a white dwarf.

Neutron stars are about the size of a city with radii of around
$12\,\mathrm{km}$ and masses up to an observed $2.01\,M_\odot$ [Antoniadis *et al.*
(2013)]. This suggests these objects have extreme central densities in
the order of $10^{15}\,\mathrm{g/cm^3}$, greater than the density of an atomic nucleus
which is about $2.5 \times 10^{14}\,\mathrm{g/cm^3}$. Newly formed NSs, known as proto-
neutron stars, can have temperatures in the order of $10^{12}\,\mathrm{K}$, but
escaping neutrinos cool these stars very quickly to around $10^{10}\,\mathrm{K}$. NSs
may rotate rapidly due to the conservation of the angular momen-
tum of the massive progenitor star, and NSs in binary systems can
be further spun up by the accretion of matter from the companion
star and have been observed to rotate at up to 716 Hz [Hessels *et al.*
(2006)]. Most NSs are observed as pulsars, rotating NSs that emit
beams of particles and electromagnetic radiation from their magnetic poles that appear to pulse when their radiation cone sweeps
by the Earth. Pulsars can be powered by rotation, matter accretion,
and extreme magnetic fields. In the latter case some NSs, known as
magnetars, have been shown to have surface magnetic fields in the

order of 10^{14} G (for comparison, the Sun's magnetic field is in the order of 1 G) [Olausen and Kaspi (2014)]. The extreme properties of NSs are very intriguing, but they alone are not the focus of this work. Instead, our intention is to study the nature and properties of matter at supranuclear densities and temperatures that are low on the nuclear scale (i.e., $T \lesssim 1\,\text{MeV}$, where $1\,\text{MeV} \approx 10^{10}\,\text{K}$). A representation of the phase diagram of dense matter is given in Fig. 4.1. Powerful particle colliders are currently investigating the high-temperature, low-density region of the phase diagram of dense matter, reproducing conditions similar to that of the early universe. However, the low-temperature, high-density (supranuclear) region of the phase diagram is currently inaccessible to terrestrial experiments. Instead this region of the phase diagram can be probed by the observation and computational modeling of NSs. In this work we focus on

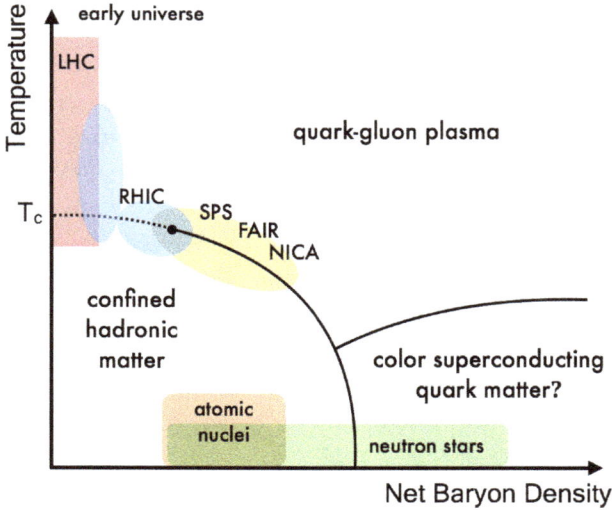

Fig. 4.1 Schematic of the phase diagram of dense matter. The particle colliders represented in the diagram are as follows: LHC is the Large Hadron Collider at CERN [LHC (2018)], RHIC is the Relativistic Heavy Ion Collider at Brookhaven National Laboratory [RHIC (2018)], SPS is the Super Proton Synchrotron at CERN [SPS (2018)], FAIR is the future Facility for Antiproton and Ion Research at GSI [FAIR (2018)], and NICA is the future Nuclotron-based Ion Collider fAcility at JINR [NICA (2018)].

the modeling of cold catalyzed NS matter and the determination of the equation of state, the relationship between energy density and pressure that is necessary for calculating the structural properties of NSs. Of particular interest is the possible presence of exotic states of matter in NS cores.

Neutron star matter is to first approximation composed of neutrons and protons that interact via the strong interaction, and leptons that must be present to ensure charge neutrality. To describe the interaction between baryons we use quantum hadrodynamics, modeling the baryon–baryon interaction as a coupling of the baryons to meson fields that reproduce the fundamental attractive and repulsive nature of the strong force. This approach is similar to the one-boson-exchange (OBE) modeling of the nuclear interaction that very successfully describes the wealth of nucleon–nucleon (NN) scattering data. However, instead of fixing the meson–nucleon coupling constants to scattering data, they are fit to the properties of bulk nuclear matter at the nuclear saturation density extracted from recent experimental or theoretical results from nuclear physics. The collection of properties utilized for this fitting procedure then constitutes a particular nuclear parameterization. This procedure can be implemented without regard for exotic baryons, as these do not typically appear in NS models until well beyond the density of nuclear matter. Extending this model to include the exotic baryons is rather straightforward. One must simply specify the meson–baryon coupling constants for the exotic baryons. Unfortunately, these range from very poorly constrained for the hyperons to completely unknown for the delta isobars. As such, an investigation of the parameter space of the meson–hyperon coupling constants constitutes a significant portion of this work.

Why would we expect exotic matter to be present in NSs? The primary constituents of NS matter, neutrons, protons, and electrons, are all fermions, particles of half-integer spin. The Pauli Exclusion Principle states that no two identical fermions can simultaneously occupy the same quantum state. Consequently, as the density of matter increases in the NS core so must the Fermi energy of the particles. Eventually the energy required to populate an additional neutron or proton may exceed that necessary to create higher mass baryons such as hyperons or delta isobars. At that point nuclear and weak reactions that exploit these additional degrees of freedom will begin to occur, minimizing the total energy and slowing the increase of the baryon chemical potentials. We refer to these higher mass hyperons and delta isobars as exotic baryons as they do not occupy stable states in terrestrial matter, or anywhere other than possibly in the cores of NSs for that matter.

Hyperons differ from neutrons and protons in that they are composed of at least one strange quark (s), and are as a result more massive. Together, the neutron (n), proton (p), and hyperons make up the spin $1/2$ baryon octet. The first hyperon (the Λ) was not discovered using a particle collider but rather by researchers at Caltech studying cosmic ray collisions with a cloud chamber. Hyperons quickly decay via the weak interaction and have a short mean lifetime in the order of 10^{-10} seconds. The idea that stable hyperons may be present in NS matter was first proposed in 1960 [Ambartsumyan (1960)]. As the NS density increases so must the nucleon Fermi momenta according to the Pauli Exclusion Principle, and hyperons may be created in reactions such as the following,

$$N + N \rightarrow N + \Lambda + K, \qquad (4.1)$$

where N represents a nucleon, either the neutron or proton, and K is a meson known as the kaon [Glendenning (1983)]. In this (and all) strong reactions strangeness S (the number of strange quarks) is conserved; here $S_\Lambda = -1$ and $S_K = +1$ making the total strangeness of the products zero. However, the produced kaon is not forbidden from decaying via the weak interaction which does not conserve quark flavor, allowing strangeness to evolve in NSs. The additional exotic baryons that may appear in NS matter are the four delta isobars ($\Delta(1232)$), the lowest mass members of the spin $3/2$ baryon decuplet. The delta isobars do not contain strange quarks like the hyperons, but are composed of only up (u) and down (d) quarks like the nucleons. However, the spin orientations of their constituent quarks are aligned such that the total spin ($J_\Delta = 3/2$) is greater than that of the nucleons ($J_N = 1/2$). In fact, the Δ^0 and Δ^+ are spin $3/2$ resonances of the nucleons, having the same quark content as the neutron and proton respectively. The delta isobars have been largely ignored in the NS literature since a seminal work by Glendenning suggested an extremely high density for their appearance thought unlikely to be reached in the NS core [Glendenning (1983)]. They are created in strong interactions, but unlike hyperons also decay via the strong interaction, and as a result have a minuscule mean lifetime in free space in the order of 10^{-24} s [Beringer (2012)]. Additional heavier baryons exist including higher spin resonances of the hyperons, but the densities reached in the NS core are not sufficient for these to be populated.

At densities of around $0.24\,\mathrm{fm}^{-3}$ baryons begin to touch [Weber (2005)], and this density, less than twice that of the nuclear density, is greatly exceeded in the cores of NSs. Baryons such as neutrons and protons are not themselves elementary particles but are each made up of three elementary particles called quarks. Quarks interact with each other and are *confined* inside baryons by this interaction, which is known as the strong interaction (or strong force). Interestingly, the strength of the strong interaction diminishes as the energy or density of matter increases, a property known as asymptotic freedom. In as early as 1965 it was suggested that as the density increases in the NS core and the boundaries of the baryons begin to overlap their constituent quarks may be freed from their confined state, resulting in a new phase of matter referred to as deconfined quark matter (or simply quark matter) [Ivanenko and Kurdgelaidze (1965)]. In addition, it has been shown that deconfined quarks are likely to form a color superconductor at low temperature (see for example [Alford *et al.* (2008)]), though it is not necessary to directly account for color superconductivity unless one is performing a thermal evolution calculation [Weber (2005)]. Modeling hadronic matter using the relativistic mean-field (RMF) approximation and deconfined quark matter using a nonlocal variant of the three-flavor Nambu–Jona–Lasinio model (n3NJL), it was found in [Orsaria *et al.* (2014)] that NSs containing hyperons, delta isobars, and deconfined quarks are capable of satisfying the $2.01\,M_\odot$ maximum mass constraint set by PSR J0348+0432 [Antoniadis *et al.* (2013)].

Little empirical evidence exists to constrain the meson–hyperon coupling constants. They are typically specified using a combination of hypernuclear potential depths in nuclear matter deduced from scarce experimental data on hypernuclei, and theoretical arguments involving quark flavor symmetries. In this work we systematically investigate the parameter space of the meson–hyperon coupling constants. We calculate the maximum mass of neutron stars in the coupling space and use the observed mass of PSR J0348+0432 to constrain the coupling constants. The meson–hyperon coupling space has previously been investigated by Weissenborn *et al.* in a number of works [Weissenborn, Chatterjee, and Schaffner-Bielich (2012a, 2013, 2014)], and more recently in [Lim, Lee, and, Oh (2018); Fortin *et al.* (2017)].

The organization of this work is as follows. In Sec. 4.2 we introduce the relativistic mean-field approximation for determining the equation of state of hadronic matter along with a number of associated nuclear parameterizations. Constraints on the NS equation of state from nuclear physics and NS observations are discussed and used to eliminate several popular nuclear parameterizations in Sec. 4.3. In Sec. 4.4 we determine the hadronic equation of state with the inclusion of hyperons and systematically investigate the parameter space of the meson–hyperon coupling constants.

4.2. Neutron Star Equation of State and Structure

The observable structural properties of NSs such as the mass and radius are determined by the NS equation of state (EoS), the relationship between pressure and energy density in a NS. To determine the EoS we must first choose a theoretical framework in which to work, and then make an ansatz for the interior composition. Then, armed with the EoS we can determine NS properties and compare to observations to see if our model is valid.

Neutron stars are born at extreme temperatures in the order of a trillion degrees ($\sim 10^{12}$ K), making them initially opaque to neutrinos. After about twenty seconds the mean free path of neutrinos grows larger than the NS radius allowing them to escape, and the temperature quickly drops shutting down the mechanisms capable of significantly changing the composition. It is therefore both sufficient and convenient to model NS matter as matter at zero temperature and in the ground state.

The interaction between the baryons that compose nuclear matter is at the fundamental level due to the interactions between their constituent quarks that are described by quantum chromodynamics (QCD). Unfortunately, it is currently intractable to model dense baryonic matter using this fundamental theory. Instead we use quantum hadrodynamics, an effective field theory that is meant to reproduce the observed behavior of baryonic matter that is in actuality due to the underlying physics of QCD.

In practice, NSs are usually broken up into two distinct regions due to a difference in composition. The NS crust is a surficial layer with a thickness of less than two kilometers that is expected to be composed of atomic nuclei, electrons, and a neutron superfluid above the neutron drip density. Beyond the crust is the core which is composed of free baryons that may include hyperons and delta isobars, and also electrons and muons. In Sec. 4.2.1 we briefly discuss the EoS of the crust which is taken from the literature. Next in Sec. 4.2.2 we discuss the relativistic mean-field approach that is used exclusively to model hadronic NS matter in this work. Finally, in Sec. 4.2.3 we outline the determination of the structural properties of NSs from the EoS.

4.2.1. *The neutron star crust*

The NS crust is defined as the region from the surface, $\mathcal{E}_{\text{surface}} \approx 10^4\,\text{g cm}^{-3}$, to a density of around $10^{14}\,\text{g cm}^{-3}$ [Weber (1999); Haensel, Potekhin, and Yakovlev (2007)]. To model the crust it is split into two parts: an outer crust with $\mathcal{E}_{\text{outer}} \lesssim 10^{11}\,\text{g cm}^{-3}$ that consists of mostly beta stable nuclei immersed in an electron gas, and an inner crust with $10^{11}\,\text{g cm}^{-3} \lesssim \mathcal{E}_{\text{inner}} \lesssim 10^{14}\,\text{g cm}^{-3}$ that consists of neutron rich nuclei and a neutron superfluid above neutron drip density. In this work we employ the EoS of Haensel and Pichon (1994) and Douchin and Haensel (2001) to model the outer and inner crusts respectively, the inner crust EoS being very similar to the Baym–Pethick–Sutherland (1971) EoS previously employed [Baym, Pethick, and Sutherland (1971); Douchin and Haensel (2001); Haensel and Pichon (1994)]. However, it should be clear that most of the NS properties studied in this work (excluding NS radius) are relatively insensitive to the crust model chosen.

4.2.2. *Hadronic matter and relativistic nuclear mean-field theory*

To model hadronic NS matter we use the relativistic mean-field (RMF) approximation that describes the baryon–baryon interaction in terms of meson fields. Due to the intractability of modeling the interactions between all of the particles in a NS the meson field strengths are taken to be equal to their mean values in the RMF approximation. These mesons include a scalar meson (σ) that describes attraction between baryons, a vector meson (ω) that describes repulsion, and an isovector meson (ρ) that is necessary for describing baryon–baryon interactions in isospin asymmetric systems. The pion, which plays a primary role in the long range description of the baryon–baryon interaction, has odd parity and as a result vanishes in the RMF approximation [Glendenning (1997)]. Additional mesons exist that describe interactions between baryons that possess strange quarks (hyperons), however, due to both the limited availability of empirical data to constrain the coupling constants of these mesons and their exclusion from sophisticated RMF parameterizations, they are not considered in this work.

The baryons B that may populate neutron star matter include the spin $\frac{1}{2}$ baryon octet comprised of the nucleons $N \in \{n, p\}$ and hyperons $Y \in \{\Lambda, \Sigma^+, \Sigma^0, \Sigma^-, \Xi^0, \Xi^-\}$, as well as the spin $\frac{3}{2}$ delta isobar quartet $\Delta(1232) \in \{\Delta^{++}, \Delta^+, \Delta^0, \Delta^-\}$, giving $B \in \{N, Y, \Delta\}$. The Lagrangian for interacting baryons is given by [Weber (1999); Glendenning (1997)]

$$\mathcal{L}_B = \sum_B \overline{\psi}_B \left[\gamma_\mu (i\partial^\mu - g_{\omega B}(n)\omega^\mu - \frac{1}{2} g_{\rho B}(n)\boldsymbol{\tau} \cdot \boldsymbol{\rho}^\mu) \right.$$
$$\left. - (m_B - g_{\sigma B}(n)\sigma) \right] \psi_B, \qquad (4.2)$$

where $g_{\sigma B}(n)$, $g_{\omega B}(n)$, and $g_{\rho B}(n)$ are the meson–baryon coupling constants, n is the baryon number density, $\boldsymbol{\tau} = (\tau_1, \tau_2, \tau_3)$ are the Pauli isospin matrices,

$$\tau_1 = \begin{pmatrix} 0 & 1 \\ 1 & 0 \end{pmatrix}, \quad \tau_2 = \begin{pmatrix} 0 & -i \\ i & 0 \end{pmatrix}, \quad \tau_3 = \begin{pmatrix} 1 & 0 \\ 0 & -1 \end{pmatrix}, \tag{4.3}$$

γ^μ are the Dirac matrices,

$$\gamma^\mu = (\gamma^0, \gamma^1, \gamma^2, \gamma^3), \quad \gamma^0 = \beta, \quad \gamma^i = \beta\alpha_i, \tag{4.4}$$

and $\boldsymbol{\alpha}$ and β are given by

$$\boldsymbol{\alpha} = \begin{pmatrix} 0 & \boldsymbol{\tau} \\ \boldsymbol{\tau} & 0 \end{pmatrix}, \quad \beta = \begin{pmatrix} I & 0 \\ 0 & -I \end{pmatrix}. \tag{4.5}$$

In the standard RMF approach the meson–nucleon coupling constants are set so as to reproduce the properties of isospin symmetric nuclear matter (SNM) at the nuclear saturation density (n_0) determined by the given parameterization (see Sec. 4.2.2.2), and are fixed to their n_0 values,

$$g_{iN}(n) = g_{iN}(n_0), \quad i \in \{\sigma, \omega, \rho\}. \tag{4.6}$$

The coupling constants of the hyperons are specified relative to those of the nucleons and will be discussed in Sec. 4.4. The Lagrangian for the leptons $\lambda \in \{e^-, \mu^-\}$ is given by

$$\mathcal{L}_{\rm L} = \sum_\lambda \overline{\psi}_\lambda (i\gamma_\mu \partial^\mu - m_\lambda)\psi_\lambda. \tag{4.7}$$

The meson Lagrangian is given by

$$\mathcal{L}_{\rm M} = \frac{1}{2}(\partial_\mu \sigma \partial^\mu \sigma - m_\sigma^2 \sigma^2) - \frac{1}{4}\omega_{\mu\nu}\omega^{\mu\nu} + \frac{1}{2}m_\omega^2 \omega_\mu \omega^\mu$$
$$+ \frac{1}{2}m_\rho^2 \boldsymbol{\rho}_\mu \cdot \boldsymbol{\rho}^\mu - \frac{1}{4}\boldsymbol{\rho}_{\mu\nu} \cdot \boldsymbol{\rho}^{\mu\nu}, \tag{4.8}$$

where $\omega_{\mu\nu} = \partial_\mu \omega_\nu - \partial_\nu \omega_\mu$ and $\boldsymbol{\rho}_{\mu\nu} = \partial_\mu \boldsymbol{\rho}_\nu - \partial_\nu \boldsymbol{\rho}_\mu$. In order for the standard RMF model to reproduce empirical values for the nuclear incompressibility (K_0) and effective nucleon mass (m^*/m_N) at n_0

additional nonlinear scalar self-interactions are necessary and contribute to the Lagrangian [Boguta and Bodmer (1977)],

$$\mathcal{L}_{\mathrm{NL}\sigma} = -\frac{1}{3}b_\sigma m_N[g_{\sigma N}(n)\sigma]^3 - \frac{1}{4}c_\sigma[g_{\sigma N}(n)\sigma]^4, \tag{4.9}$$

where b_σ and c_σ are constants fixed by the SNM parameterization. Additionally, in order to soften the high density EoS to satisfy nuclear matter constraints (see Sec. 4.3.1.2) and reproduce empirical values for the slope of the asymmetry energy (L_0) at n_0 the following nonlinear vector self-interactions and cross-interactions can be introduced [Horowitz and Piekarewicz (2001b,a); Müeller and Serot (1996)],

$$\mathcal{L}_{\mathrm{NL}\omega} = \frac{1}{4}g_{\omega^4}(g_{\omega N}^2\omega_\mu\omega^\mu)^2, \tag{4.10}$$

$$\mathcal{L}_{\sigma\omega\rho} = \frac{1}{2}g_{\rho N}^2\boldsymbol{\rho}_\nu \cdot \boldsymbol{\rho}^\nu(g_{\sigma\rho}g_{\sigma N}^2\sigma^2 + g_{\omega\rho}g_{\omega N}^2\omega_\mu\omega^\mu), \tag{4.11}$$

where g_{ω^4}, $g_{\sigma\rho}$, and $g_{\omega\rho}$ are coupling constants fixed by the SNM parameterization. The final standard RMF Lagrangian is then given by

$$\mathcal{L}_{\mathrm{RMF}} = \mathcal{L}_{\mathrm{B}} + \mathcal{L}_{\mathrm{L}} + \mathcal{L}_{\mathrm{M}} + \mathcal{L}_{\mathrm{NL}\sigma} + \mathcal{L}_{\mathrm{NL}\omega} + \mathcal{L}_{\sigma\omega\rho}. \tag{4.12}$$

The baryon and meson field equations are obtained by evaluating the Euler–Lagrange equations for the fields in $\mathcal{L}_{\mathrm{RMF}}$ giving the following,

$$(i\gamma^\mu\partial_\mu - m_B)\psi_B = \left[g_{\omega B}(n)\gamma^\mu\omega_\mu + \frac{1}{2}g_{\rho B}(n)\gamma^\mu\boldsymbol{\tau}\cdot\boldsymbol{\rho_\mu}\right.$$

$$\left. - g_{\sigma B}(n)\sigma\right]\psi_B, \tag{4.13}$$

$$(\partial^\mu\partial_\mu + m_\sigma^2)\sigma = \sum_B g_{\sigma B}(n)\overline{\psi}_B\psi_B - b_\sigma m_n g_{\sigma N}(n)[g_{\sigma N}(n)\sigma]^2$$

$$- c_\sigma g_{\sigma N}(n)[g_{\sigma N}(n)\sigma]^3$$

$$- g_{\sigma\rho}(n)g_{\sigma N}^2(n)\sigma g_{\rho N}^2(n)\rho^2, \tag{4.14}$$

$$\partial^\mu \omega_{\mu\nu} + m_\omega^2 \omega_\nu = \sum_B g_{\omega B}(n)\overline{\psi}_B \gamma_\nu \psi_B - g_{\omega 4} g_{\omega N}(n)[g_{\omega N}(n)\omega]^3$$

$$- g_{\omega\rho}(n)g_{\omega N}^2(n)\omega g_{\rho N}^2(n)\rho^2, \tag{4.15}$$

$$\partial^\mu \boldsymbol{\rho}_{\mu\nu} + m_\rho^2 \boldsymbol{\rho}_\nu = \sum_B g_{\rho B}(n)\overline{\psi}_B \boldsymbol{\tau} \gamma_\nu \psi_B - g_{\sigma\rho}(n)g_{\sigma N}^2(n)\sigma^2 g_{\rho N}^2(n)\rho$$

$$- g_{\omega\rho}(n)g_{\omega N}^2(n)\omega^2 g_{\rho N}^2(n)\rho. \tag{4.16}$$

To simplify the simultaneous solution of these field equations the relativistic mean-field approximation is applied to the system under the assumption that the matter of interest is static uniform matter in the ground state [Glendenning (1997)]. The meson fields and baryon currents are replaced by their mean values and the field equations (4.14)–(4.16) are replaced by the meson mean-field equations,

$$m_\sigma^2 \bar{\sigma} = \sum_B g_{\sigma B}(n)\langle\overline{\psi}_B \psi_B\rangle - b_\sigma m_N g_{\sigma N}(n)\left[g_{\sigma N}(n)\bar{\sigma}\right]^2$$

$$- c_\sigma g_{\sigma N}(n)\left[g_{\sigma N}(n)\bar{\sigma}\right]^3 - g_{\sigma\rho}g_{\sigma N}^2 \bar{\sigma} g_{\rho N}^2 \bar{\rho}^2 \tag{4.17}$$

$$m_\omega^2 \bar{\omega} = \sum_B g_{\omega B}(n)\langle\psi_B^\dagger \psi_B\rangle - g_{\omega 4} g_{\omega N}(n)\left[g_{\omega N}(n)\bar{\omega}\right]^3$$

$$- g_{\omega\rho}(n)g_{\omega N}^2(n)\bar{\omega} g_{\rho N}^2(n)\bar{\rho}^2, \tag{4.18}$$

$$m_\rho^2 \bar{\rho} = \sum_B g_{\rho B}(n)I_{3B}\langle\psi_B^\dagger \psi_B\rangle - g_{\sigma\rho}(n)g_{\sigma N}^2(n)\bar{\sigma}^2 g_{\rho N}^2(n)\bar{\rho}$$

$$- g_{\omega\rho}(n)g_{\omega N}^2(n)\bar{\omega}^2 g_{\rho N}^2(n)\bar{\rho}. \tag{4.19}$$

where $\bar{\sigma} \equiv \langle\sigma\rangle$, $\bar{\omega} \equiv \langle\omega_0\rangle$, and $\bar{\rho} \equiv \langle\rho_0\rangle$. The expectation values of the baryon currents in the mean-field equations are the baryon scalar density,

$$n_B^S \equiv \langle\overline{\psi}_B \psi_B\rangle = \frac{2J_B + 1}{2\pi^2} \int_0^{k_B} \frac{m_B^*(\bar{\sigma})}{\sqrt{k^2 + m_B^{*2}(\bar{\sigma})}} k^2 \, dk$$

$$= \sum_B \frac{2J_B + 1}{2\pi^2} \left(\frac{m_B^*}{2}\right) \left[k_B E_B - m_B^{*2} \ln\left(\frac{E_B + k_B}{m_B^*}\right)\right], \tag{4.20}$$

and the baryon number (vector) density,

$$n_B \equiv \langle \psi_B^\dagger \psi_B \rangle = \frac{2J_B + 1}{6\pi^2} k_B^3, \tag{4.21}$$

where J_B is the spin, k_B is the Fermi momentum,

$$m_B^*(\bar{\sigma}) = m_B - g_{\sigma B}(n)\bar{\sigma} \tag{4.22}$$

is the effective mass, and $E_B = \sqrt{k_B^2 + m_B^{*2}}$ is the Fermi energy of baryon species B.

To determine the energy density and pressure at a given baryon number density the meson mean-field equations must first be solved subject to the condition that NS matter be charge neutral,

$$\sum_B n_B q_B + \sum_\lambda n_\lambda q_\lambda = 0, \tag{4.23}$$

where q_B and q_λ are the baryon and lepton electric charge in units of the elementary charge, and that baryon number be conserved,

$$n - \sum_B n_B = 0. \tag{4.24}$$

This constitutes a system of five coupled nonlinear equations that are solved simultaneously to determine the meson mean-fields ($\bar{\sigma}$, $\bar{\omega}$, $\bar{\rho}$) and the neutron and electron Fermi momenta (k_n, k_e). The Fermi momenta of the rest of the baryons are governed by the condition that NS matter be in chemical equilibrium,

$$\mu_B = \mu_n - q_B \mu_e, \tag{4.25}$$

where μ_B is the baryon chemical potential given by

$$\mu_B = g_{\omega B}(n)\bar{\omega} + g_{\rho B}(n)I_{3B}\bar{\rho} + \sqrt{k_B^2 + m_B^{*2}}, \tag{4.26}$$

and $\mu_e = E_e$ is the electron chemical potential. When $\mu_B < \mu_n - q_B \mu_e$ it is energetically favorable for the system to populate baryon state B. Thus the inclusion of additional baryonic degrees of freedom (hyperons or Δs) softens the equation of state by reducing the overall chemical potential, lowering the Fermi momenta of the baryons in the system and minimizing the total energy. Additional leptons are also

Table 4.1   Properties of the nucleons and leptons that appear in NSs. Given are the mass m, electric charge q, the total spin J, the z component of the isospin I_3 (nucleons only), and the constituent quarks (nucleons only). The masses are those published by the Particle Data Group [Patrignani *et al.* (2016)].

Particle	m (MeV)	q (e)	J	I_3	Quarks
n	939.565	0	1/2	$-1/2$	udd
p	938.272	$+1$	1/2	1/2	uud
e^-	0.510999	-1	1/2	—	—
μ^-	105.658	-1	1/2	—	—

possible though are typically limited to the muon which is populated when $\mu_\mu < \mu_e$. The properties of the nucleons and leptons are given in Table 4.1, while the hyperon properties are tabulated in Sec. 4.4.

To calculate the RMF energy density ($\mathcal{E}$) and pressure (P) one must first determine the energy–momentum tensor, which is calculated from the metric ($g_{\mu\nu}$) and the RMF Lagrangian as follows,

$$T^{\mu\nu} = -g^{\mu\nu}\mathcal{L} + \sum_\phi \frac{\partial \mathcal{L}}{\partial\left(\partial_\mu \phi\right)}\partial^\nu \phi, \qquad (4.27)$$

where $\phi \in \{\sigma, \omega, \rho\}$. Spacetime is to a very good approximation flat on the length scale of particle interactions in NS matter ($\sim 1\,\mathrm{fm}$), so the metric is simply the flat space Minkowski metric, $g_{\mu\nu} = \mathrm{diag}(-1, 1, 1, 1)$. Finally, the expressions for the energy density and pressure are as follows,

$$\mathcal{E}_{\mathrm{RMF}} = \langle T_{00}\rangle$$

$$= \sum_B \frac{2J_B + 1}{2\pi^2}\int_0^{k_B}\sqrt{k^2 + m_B^{*2}(\bar{\sigma})}\, k^2 dk$$

$$+ \sum_\lambda \frac{1}{\pi^2}\int_0^{k_\lambda}\sqrt{k^2 + m_\lambda^2}\, k^2 dk$$

$$+ \frac{1}{2}m_\sigma^2\bar{\sigma}^2 + \frac{1}{2}m_\omega^2\bar{\omega}^2 + \frac{1}{2}m_\rho^2\bar{\rho}^2 - \mathcal{L}_{\mathrm{NL}\sigma} + 3(\mathcal{L}_{\mathrm{NL}\omega} + \mathcal{L}_{\sigma\omega\rho}),$$

$$(4.28)$$

$$
\begin{aligned}
P_{\text{RMF}} &= \frac{1}{3} \sum_i \langle T_{ii} \rangle \\
&= \frac{1}{3} \sum_B \frac{2J_B + 1}{2\pi^2} \int_0^{k_B} \frac{k^4 dk}{\sqrt{k^2 + m_B^{*2}(\bar{\sigma})}} \\
&+ \frac{1}{3} \sum_\lambda \frac{1}{\pi^2} \int_0^{k_\lambda} \frac{k^4 dk}{\sqrt{k^2 + m_\lambda^2}} \\
&- \frac{1}{2} m_\sigma^2 \bar{\sigma}^2 + \frac{1}{2} m_\omega^2 \bar{\omega}^2 + \frac{1}{2} m_\rho^2 \bar{\rho}^2 + \mathcal{L}_{\text{NL}\sigma} + \mathcal{L}_{\text{NL}\omega} + \mathcal{L}_{\sigma\omega\rho}.
\end{aligned}
\tag{4.29}
$$

4.2.2.1. *Density-dependent RMF (DDRMF)*

The density-dependent relativistic mean-field (DDRMF) model for the hadronic EoS (often referred to as DDRH for density-dependent relativistic Hartree) is an extension of the standard RMF approach that attempts to account for medium effects by making the meson–baryon coupling constants dependent on the local baryon number density [Fuchs, Lenske, and Wolter (1995)]. This density dependence is typically extracted from properties of nuclear matter or finite nuclei, solidifying the model empirically to a greater degree than in standard RMF.

The meson–baryon coupling constants, constant in the case of standard RMF (see (4.6)), are taken to be dependent on the baryon number density as follows,

$$
g_{iB}(n) = g_{iB}(n_0) f_i(x), \tag{4.30}
$$

where $i \in \{\sigma, \omega, \rho\}$, $x = n/n_0$, and $f_i(x)$ provides the functional form for the density dependence. The most commonly utilized ansatz for $f_i(x)$ are given by [Typel and Wolter (1999)]

$$
f_i(x) = a_i \frac{1 + b_i(x + d_i)^2}{1 + c_i(x + d_i)^2}, \tag{4.31}
$$

for $i \in \{\sigma, \omega\}$, and

$$
f_\rho(x) = \exp[-a_\rho(x - 1)]. \tag{4.32}
$$

The nine parameters of the density dependence $(a_\sigma, b_\sigma, c_\sigma, d_\sigma, a_\omega, b_\omega,$ $c_\omega, d_\omega, a_\rho)$, the values of the meson–nucleon couplings at n_0 $(g_{\sigma N}(n_0),$ $g_{\omega N}(n_0), g_{\rho B}(n_0))$, and the mass of the scalar meson (m_σ) are all fit to properties of SNM at n_0 and to the properties of finite nuclei including but not limited to binding energies, charge and diffraction radii, spin–orbit splittings, and neutron skin thickness (see [Lalazissis *et al.* (2005); Typel (2005)]).

The density dependence of the meson–baryon couplings eliminates the need for the nonlinear self-interactions and cross-interactions in (4.9)–(4.11), modifying the RMF Lagrangian (4.12) as follows,

$$\mathcal{L}_{\text{DDRMF}} = \mathcal{L}_{\text{RMF}} - \mathcal{L}_{\text{NL}\sigma} - \mathcal{L}_{\text{NL}\omega} - \mathcal{L}_{\sigma\omega\rho}. \tag{4.33}$$

Applying the Euler–Lagrange equations to the DDRMF Lagrangian leads to a rearrangement contribution (Σ_r) to the chemical potential of the baryons not present in standard RMF [Fuchs, Lenske, and Wolter (1995)],

$$\Sigma_r = \sum_B \left(\frac{\partial g_{\omega B}}{\partial n} n_B \bar{\omega} + \frac{\partial g_{\rho B}}{\partial n} I_{3B} n_B \bar{\rho} - \frac{\partial g_{\sigma B}}{\partial n} n_B^S \bar{\sigma} \right). \tag{4.34}$$

The rearrangement energy contributes to the baryon chemical potential as follows,

$$\mu_B = g_{\omega B}(n)\bar{\omega} + g_{\rho B}(n) I_{3B}\bar{\rho} + \sqrt{k_B^2 + m_B^{*2}} + \Sigma_r, \tag{4.35}$$

effectively coupling the number density and Fermi momenta of all the baryons, negating the simple relationship between the two given by (4.21). Consequently, the standard RMF nonlinear system must be modified to include a separate chemical equilibrium equation (4.25) for each baryon (excluding the neutron).

Finally, the energy density and pressure are also modified by the removal of the nonlinear self-interactions and cross-interactions, and the rearrangement energy contributes to the pressure as follows,

$$\mathcal{E}_{\text{DDRMF}} = \mathcal{E}_{\text{RMF}} - \mathcal{L}_{\text{NL}\sigma} - 3(\mathcal{L}_{\text{NL}\omega} + \mathcal{L}_{\sigma\omega\rho}), \tag{4.36}$$

$$P_{\text{DDRMF}} = P_{\text{RMF}} + \mathcal{L}_{\text{NL}\sigma} - \mathcal{L}_{\text{NL}\omega} - \mathcal{L}_{\sigma\omega\rho} + n\Sigma_r. \tag{4.37}$$

4.2.2.2. *The RMF and DDRMF parameterizations*

Neutron stars generally have a much larger number of neutrons $(I_{3,n} = -\frac{1}{2})$ than protons $(I_{3,p} = +\frac{1}{2})$, $n_n \gg n_p$, making them highly isospin asymmetric systems. However, an EoS of isospin symmetric nuclear matter (SNM) can be constructed by requiring that $n_n = n_p$ and removing the leptons, or simply relaxing the condition of charge neutrality, and this approximation can be parameterized to reproduce empirically determined properties of nuclear matter at saturation density, thereby constraining the EoS. These saturation properties are the binding energy per nucleon (E_0), the nuclear incompressibility (K_0), the isospin asymmetry energy (J) and its slope (L_0), and the effective mass (m^*/m_N), all computed at the nuclear saturation density (n_0). The saturation properties associated with the RMF and DDRMF parameterizations employed in this work are given in Tables 4.2 and 4.3 respectively.

The EoS of SNM for the standard RMF model is fixed to the properties of nuclear matter at saturation density by five parameters: the meson–nucleon coupling constants $g_{\sigma N}(n_0)$, $g_{\omega N}(n_0)$, and

Table 4.2 Properties of SNM at saturation density for RMF(L) parameterizations. Properties include the nuclear saturation density n_0, energy per nucleon E_0, nuclear incompressibility K_0, effective nucleon mass m^*/m_N, asymmetry energy J, asymmetry energy slope L_0, and nucleon potential U_N. The SNM parameterization acronyms are given as follows: Spinella and Weber (SW), Glendenning and Moszkowski (GM1, GM3), NonLinear 3 (NL3), and Indian University–Florida State University (IUFSU).

Saturation Property	SW(L)	GM1(L)	GM3(L)	NL3	IUFSU
n_0 (fm^{-3})	0.150	0.153	0.153	0.148	0.155
E_0 (MeV)	-16.0	-16.3	-16.3	-16.3	-16.4
K_0 (MeV)	260.0	300.0	240.0	271.8	231.33
m^*/m_N	0.70	0.70	0.78	0.60	0.61
J (MeV)	31.0	32.5	32.5	37.4	31.3
L_0 (MeV)	89.9(55.0)	94.0(55.0)	89.7(55.0)	118.5	47.2
$-U_N$ (MeV)	64.6	65.5	60.8	71.8	72.8
Reference		[Glendenning (1991)]	[Glendenning (1991)]	[Lalazissis (1997)]	[Fattoyev (2010)]

Table 4.3   Properties of SNM at saturation density for DDRMF parameterizations. Properties include the nuclear saturation density n_0, energy per nucleon E_0, nuclear incompressibility K_0, effective nucleon mass m^*/m_N, asymmetry energy J, asymmetry energy slope L_0, and nucleon potential U_N. The SNM parameterization acronyms are given as follows: Typel and Wolter 1999 (TW99), Density Dependent 2 (DD2), Density Dependent F (DDF), and Meson Exchange 2 (ME2).

Saturation Property	TW99	DD2	DDF	ME2
n_0 (fm^{-3})	0.153	0.149	0.147	0.152
E_0 (MeV)	-16.25	-16.02	-16.04	-16.14
K_0 (MeV)	240.27	242.72	223.32	250.92
m^*/m_N	0.55	0.56	0.56	0.57
J (MeV)	32.77	31.67	31.63	32.30
L_0 (MeV)	55.31	55.04	56.00	51.25
$-U_N$ (MeV)	77.1	75.2	75.3	75.1
Reference	[Typel (1999)]	[Typel (2010)]	[Klahn (2006)]	[Lalazissis (2005)]

$g_{\rho N}(n_0)$, and the coefficients of the nonlinear scalar self-interactions b_σ and c_σ. These parameters can be fixed to the desired saturation properties algebraically as shown in [Glendenning (1997)]. In order to also fix the slope of the asymmetry energy at saturation density one must go beyond the standard RMF approach, as will be discussed in the following section. The parameters associated with the saturation properties of the RMF parameterizations given in Table 4.2 are listed in Table 4.4.

The additional nine parameters required for the functional density dependence of the meson–nucleon coupling constants in the DDRMF approach are typically fit to reproduce empirically determined properties of finite nuclei (usually doubly magic nuclei) as described in Sec. 4.2.2.1 [Lalazissis *et al.* (2005); Typel (2005)]. These parameters are listed in Table 4.5 for the saturation properties given in Table 4.3.

The properties of SNM at saturation density and the available constraints will be discussed in further detail in Sec. 4.3.

Table 4.4   Parameters of RMF(L) models that produce the properties of SNM at saturation density given in Table 4.2.

Parameters	SW(L)	GM1(L)	GM3(L)	NL3	IUFSU
m_σ	550	550	550	508.194	491.5
m_ω	783	783	783	782.501	782.5
m_ρ	763	770	770	763	763
$g_{\sigma N}$	9.7744	9.5722	8.7844	10.176	9.9713
$g_{\omega N}$	10.746	10.618	8.7195	12.789	13.032
$g_{\rho N}$	7.8764	8.1983	8.5441	8.9853	13.590
$100\,b_\sigma$	0.3798	0.2936	0.8629	0.2099	0.1800
$100\,c_\sigma$	-0.3197	-0.1068	-0.2434	-0.2668	0.0049
$g_{\omega 4}$	—	—	—	—	0.005
$g_{\sigma \rho}$	—	—	—	—	0.0
$g_{\omega \rho}$	—	—	—	—	0.092
a_ρ	(0.3796)	(0.3898)	(0.3199)	—	—
Reference		[Glendenning (1991)]	[Glendenning (1991)]	[Lalazissis (1997)]	[Fattoyev (2010)]

Table 4.5   Parameters of DDRMF models that produce the properties of SNM at saturation density given in Table 4.3.

Parameters	TW99	DD2	DDF	ME2
m_σ	550	546.21246	555	550.1238
m_ω	783	783	783	783
m_ρ	763	763	763	763
$g_{\sigma N}$	10.729	10.687	11.024	10.540
$g_{\omega N}$	13.290	13.342	13.575	13.019
$g_{\rho N}$	3.6610	3.6269	3.6450	3.6838
a_σ	1.365469	1.357630	1.4867	1.3881
b_σ	0.226061	0.634442	0.19560	1.0943
c_σ	0.409704	1.005358	0.42817	1.7057
d_σ	0.901995	0.575810	0.88233	0.4421
a_ω	1.402488	1.369718	1.5449	1.3892
b_ω	0.172577	0.496475	0.18381	0.9240
c_ω	0.344293	0.817753	0.43969	1.4620
d_ω	0.983955	0.638452	0.87070	0.4775
a_ρ	0.515	0.518903	0.44793	0.5647
Reference	[Typel (1999)]	[Typel (2010)]	[Klahn (2006)]	[Lalazissis (2005)]

4.2.2.3. *RMF with a density-dependent isovector meson–baryon coupling constant (RMFL)*

The standard RMF approach is parameterized to reproduce a number of properties of SNM at n_0 including the asymmetry energy $J = E_{\mathrm{sym}}(n_0)$ (see Table 4.2). However, the standard RMF approximation is not parameterized to fix the slope of the asymmetry energy (L_0) at n_0, a quantity that has become more tightly constrained in recent years and may have a significant effect on NS composition and properties [Cavagnoli, Menezes, and Providência, (2011); Danielewicz and Lee (2014); Lattimer and Lim (2013); Lattimer and Steiner (2014); Providência and Rabhi (2013); Providência *et al.* (2014)]. The additional parameters that scale the nonlinear vector self-interaction and vector–isovector cross interaction contributions to the Lagrangian can be set to satisfy these constraints on L_0. However, parameterizations that include these contributions, such as the IUFSU parameterization, typically produce fairly soft EoS that cannot satisfy the maximum NS mass constraint (see Sec. 4.3.2.1) when hyperons are accounted for in the composition.

Instead of including additional nonlinear contributions to the Lagrangian that soften the EoS it is also possible to augment the standard RMF approach by making the isovector meson–baryon coupling constant density-dependent as in the DDRMF approach,

$$g_{\rho B}(n) = g_{\rho B}(n_0) \exp[-a_\rho(x - 1)]. \qquad (4.38)$$

This is a very convenient method as L_0 can be tailored by adjusting the coefficient a_ρ without affecting the other saturation properties, essentially leaving the preexisting core parameterization intact [Drago *et al.* (2014a)]. In this work we refer to this approach as RMFL. The density dependence of $g_{\rho B}$ does complicate the nonlinear system in the same fashion as DDRMF, as a rearrangement energy contribution must be accounted for,

$$\Sigma_r = \sum_B \frac{\partial g_{\rho B}}{\partial n} I_{3B} n_B \bar{\rho}. \qquad (4.39)$$

The SWL, GM1L, and GM3L RMFL parameterizations are identified in Table 4.2 as SW(L), GM1(L), and GM3(L), as the only difference in the saturation properties between these parameterizations and their standard RMF counterparts is the value of L_0. In each of these three parameterizations we have fixed $L_0 = 55\,\text{MeV}$, and the associated values of the coefficient a_ρ are given in Table 4.4. The standard RMF NL3 parameterization was neglected because the core parameterization suffers from an extremely high asymmetry energy that is inconsistent with current constraints (see Sec. 4.3.1.1).

4.2.3. *Neutron star structure*

The gravitational mass M and radius R of a spherical nonrotating NS can be determined by solving a set of coupled differential stellar structure equations known as the Tolman–Oppenheimer–Volkoff (TOV) equation [Oppenheimer and Volkoff (1939); Tolman (1939)],

$$\frac{dP}{dr} = -\frac{[\mathcal{E}(r) + P(r)][m(r) + 4\pi r^3 P(r)]}{r^2[1 - 2m(r)/r]}, \tag{4.40}$$

$$\frac{dm}{dr} = 4\pi r^2 \mathcal{E}(r), \tag{4.41}$$

where r is the radial distance from the NS center and $G = c = 1$.

It is also convenient to determine the total baryon number N_B of the NS by simultaneously integrating the following along with the TOV equation,

$$\frac{dA}{dr} = \frac{4\pi r^2 n(r)}{\sqrt{1 - 2m(r)/r}}, \tag{4.42}$$

with $N_B = A(R)$. This quantity is useful for determining properties such as the strange quark fraction or the baryon fraction of a particular baryon species.

By solving the TOV equation for a range of central densities one can construct a mass–radius curve that describes a family of NSs and can be used to determine properties such as the maximum possible NS mass and the radius of the canonical $1.4\,M_\odot$ NS for the given EoS model. These quantities among others can also be used to constrain and rule out EoS models and parameterizations. Constraints on the mass and radius from observations of NSs will be discussed in the next section.

4.3. Constraining the Equation of State

To investigate the composition of NSs and in particular the presence of exotic matter in the NS core it is important that we start with a model that is consistent with what we already know about NSs and dense matter. Much is known about matter at nuclear densities, as atomic nuclei have been studied extensively for more than a half a century. In this regard, RMF models are fixed to several empirical properties of matter at the nuclear saturation density, and DDRMF models are additionally parameterized to reproduce numerous properties of finite nuclei. However, the nuclear saturation density is just a single point in the NS EoS, and little is known about the EoS at supranuclear densities. To further constrain the NS EoS we must rely primarily on NS observations and the subsequent determination of macroscopic NS properties. NSs emit electromagnetic radiation through numerous mechanisms. Thermal emission from the NS surface may be detectable in the optical–UV range. Pulsars rotate and project beams of particles and radiation from the magnetic poles that are detectable if the beam happens to intersect with the Earth. NSs in binary systems may emit X-ray (or possibly even gamma ray) radiation powered by the accretion of matter from the binary companion. In addition, the observation of a binary companion can provide information about the gravitational interaction (orbit) of the system. These observations can be utilized to determine properties including but not limited to the NS mass, radius, rotational frequency, and temperature, which can in turn be used to constrain the NS EoS.

In this section we will first discuss the properties of SNM used to parameterize the RMF(L) and DDRMF models. Following that we will discuss properties of NSs that can be used to constrain the EoS. Constraints from the recent literature will be presented in both cases where available, and our models and parameterizations will be evaluated in light of those constraints.

4.3.1. *Constraining the equation of state of symmetric nuclear matter*

The standard RMF approach for modeling the EoS of NS matter is parameterized to reproduce the following properties of SNM at the nuclear saturation density (n_0): the binding energy per nucleon (E_0), the nuclear incompressibility (K_0), the isospin asymmetry energy (J), and the effective mass (m^*/m_N). Introducing a nonlinear vector–isovector cross interaction or a density-dependent isovector meson–baryon coupling constant provides an additional parameter that can be fit to the slope of the asymmetry energy (L_0) at n_0. The saturation properties and parameter values for the parameterizations considered in this work were given in Tables 4.2–4.5. In this section we will discuss the saturation properties of SNM and compare those of the chosen parameterizations to constraints from the recent literature.

4.3.1.1. *Properties of symmetric nuclear matter*

The central density of atomic nuclei is known to remain relatively constant with an increase of nucleons and can therefore be approximated by simply dividing the number of nucleons by the nuclear volume,

$$n_0 = \frac{A}{\frac{4}{3}\pi R^3}, \tag{4.43}$$

where $R = r_0 A^{1/3}$ and $r_0 \approx 1.16$ fm is an empirically determined constant [Glendenning (1997); Moller and Nix (1981)]. This (n_0) is known as the nuclear saturation density and fixes the location of zero pressure and the minimum of the binding energy per nucleon in SNM. There are numerous methods for estimating the saturation density, and it has been suggested that the results constrain the value to the range $n_0 = 0.17 \pm 0.03$ fm^{-3} [Dutra *et al.* (2012)], though RMF parameterizations place it consistently in the vicinity of $n_0 \approx 0.15$ fm^{-3}.

The binding energy per nucleon of atomic nuclei is given by the Bethe–Weizsaecker formula

$$\frac{B}{A} = a_{\text{volume}} - a_{\text{surface}} \frac{1}{A^{1/3}} - a_{\text{Coulomb}} \frac{Z(Z-1)}{A^{4/3}}$$
$$- a_{\text{symmetry}} \frac{(A - 2Z)^2}{A^2} + \cdots, \tag{4.44}$$

where A is the atomic number and Z is the proton number. The saturation density value of the energy per nucleon, $E_0 = (B/A)_{n=n_0}$, is related to the saturation energy density $(\mathcal{E}_0)$ by the following,

$$E_0 = \frac{\mathcal{E}_0}{n_0} - m_N. \qquad (4.45)$$

The coefficients a_i can be determined by fitting to experimental nuclear binding energies or to nuclear masses via the semi-empirical mass formula,

$$M(A, Z) = (A - Z)m_n + Zm_p - B. \qquad (4.46)$$

In SNM A becomes very large and the binding energy per nucleon simplifies to the volume contribution, $B/A \approx a_{\text{volume}}$, which may lie within the recently suggested range of $-15.9 \pm 0.4\,\text{MeV}$ at saturation density [Thew *et al.* (2017)].

Figure 4.2 shows the energy per nucleon for all parameterizations (excluding RMFL parameterizations which do not alter B/A of their standard RMF counterparts), with an enlarged view of the saturation region in the right panel. Only the IUFSU parameterization is inconsistent with the range in E_0 suggested above, but this will not serve to eliminate it from further analyses.

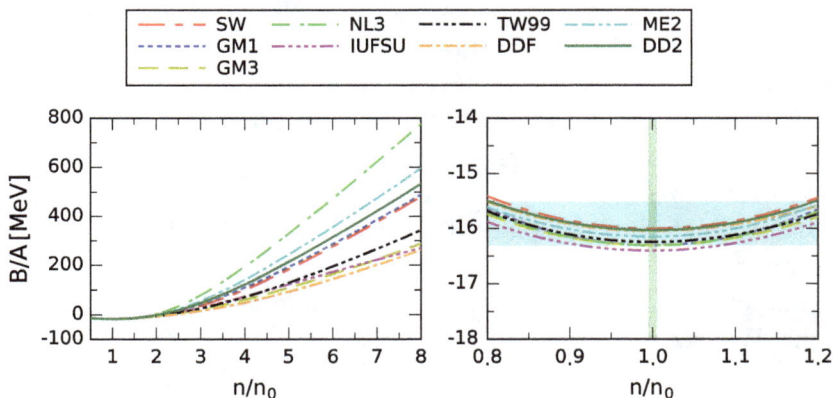

Fig. 4.2   The energy per nucleon B/A as a function of baryon number density in SNM. The right panel enlarges the region around the saturation density. The cyan shading represents the acceptable range suggested in [Kolomeitsev, Maslov, and Voskresensky (2017)], while the green shading identifies the saturation density for convenience.

The incompressibility of nuclear matter (or compression modulus) K is defined by the curvature of the EoS as follows,

$$K(n) = 9n \left(\frac{\partial^2 \mathcal{E}}{\partial n^2}\right), \tag{4.47}$$

with K_0 representing the incompressibility at saturation density, $K_0 = K(n_0)$. The incompressibility can be computed and verified by numerically solving for the EoS of SNM using a backward difference scheme given by the following (in second order accuracy),

$$K_i = 9n_i \frac{2\mathcal{E}_i - 5\mathcal{E}_{i-1} + 4\mathcal{E}_{i-2} - \mathcal{E}_{i-3}}{\Delta n^2} + O(\Delta n^2), \tag{4.48}$$

where $\mathcal{E}$ is the energy density and $\Delta n = n_i - n_{i-1}$ is held constant.

Theoretical values for K_0 are typically extracted from measurements of the giant monopole resonance (GMR), the radial vibration of nucleons in a nucleus. The most commonly utilized constraint for K_0 comes from a review of experimental values of the GMR and suggests a range of 220–260 MeV [Shlomo, Kolomietz, and Colo (2006)]. However, a more recent GMR analysis suggests the possibility of a higher range of 250–315 MeV (see [Stone, Stone, and Moszkowski (2014)] for this analysis as well as a review of previously published theoretical values of K_0). These constraints are listed in Table 4.6 and labeled K1 and K2 respectively. Figure 4.3 shows the incompressibility as a function of baryon number density with an enlarged view in the right panel. The NL3 and GM1 RMF parameterizations satisfy the higher constraint K2 but fail to satisfy the lower constraint K1. The SW RMF parameterization just satisfies both constraints by design, with a saturation incompressibility $K_0 = 260$ MeV. The rest all satisfy the lower constraint K1.

Table 4.6 Constraints on the nuclear incompressibility considered in this work. The constraints are labeled for convenience.

Reference	K_0 (MeV)	Designation
[Shlomo *et al.* (2006)]	220–260	K1
[Stone *et al.* (2014)]	250–310	K2

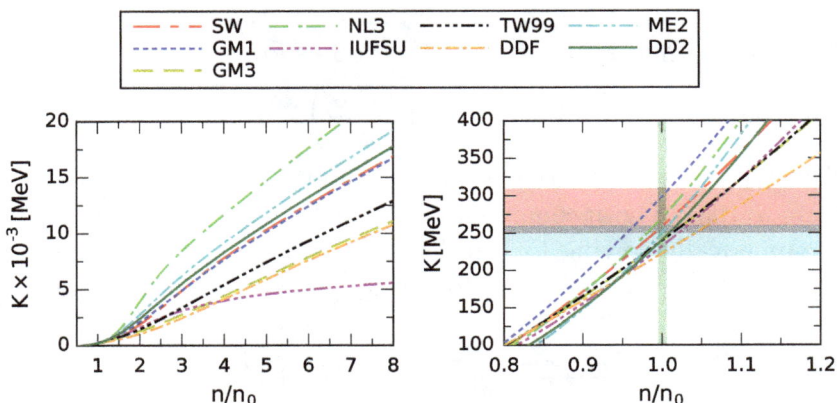

Fig. 4.3   The nuclear incompressibility K as a function of baryon number density in SNM. The right panel enlarges the region around the saturation density. The cyan shading represents the constraint K1, while the red shading indicates the constraint K2 (see Table 4.6). The green shading identifies the saturation density for convenience.

The isospin asymmetry energy of nuclear matter, represented by the proportionality constant a_{symmetry} in the binding energy per nucleon (4.44), has been shown to have a substantial effect on NS properties (see for example [Lattimer and Steiner (2014); Providência *et al.* (2014)]). The isospin asymmetry energy of SNM as a function of baryon number density is given by

$$E_{\text{sym}}(n) = \frac{1}{2}\left(\frac{\partial^2 (\mathcal{E}/n)}{\partial \delta^2}\right)_{\delta=0}, \tag{4.49}$$

where δ is the asymmetry parameter,

$$\delta = \frac{n_n - n_p}{n_n + n_p}. \tag{4.50}$$

The asymmetry energy can be expressed analytically for RMF(L) parameterizations by [Glendenning (1997); Dutra *et al.* (2014)]

$$E_{\text{sym}}(n) = \left(\frac{k_N^3}{12\pi^2}\right)\frac{g_\rho^2}{m_\rho^2 + g_{\sigma\rho}g_\sigma^2 g_\rho^2 \bar{\sigma}^2 + g_{\omega\rho}g_\omega^2 g_\rho^2 \bar{\omega}^2} + \frac{k_N^2}{6\sqrt{k_N^2 + m^{*2}}}, \tag{4.51}$$

and for DDRMF parameterizations by [Li *et al.* (2008)]

$$E_{\text{sym}}(n) = \frac{n}{2}\left(\frac{g_\rho}{m_\rho}\right)^2 + \frac{k_N^2}{6\sqrt{k_N^2 + m^{*2}}}. \tag{4.52}$$

The DDRMF expression differs from that of RMF(L) due to the presence of the rearrangement energy, as there is no longer a simple relation between the Fermi momentum and number density of the nucleons. The value of the asymmetry energy at saturation density is denoted as $J = E_{\text{sym}}(n_0)$. The slope of the asymmetry energy is given by

$$L(n) = 3n\left(\frac{\partial E_{\text{sym}}}{\partial n}\right), \tag{4.53}$$

with $L_0 = L(n_0)$. L can be computed and verified by numerically solving for the EoS using a backward difference scheme given by the following (in second order accuracy):

$$L_i = 3n_i \frac{3E_{\text{sym},i} - 4E_{\text{sym},i-1} + E_{\text{sym},i-2}}{2\Delta n} + O(\Delta n^2). \tag{4.54}$$

The asymmetry energy at saturation density can be extracted from empirical data using the semi-empirical mass formula. However, in order to constrain J and L to a given range for the studied parameterizations, the results are taken from recent literature that combine different experimental and theoretical approaches. These constraints are listed in Table 4.7. Figure 4.4 shows the asymmetry energy as a function of baryon number density (left panel) and the saturation asymmetry energy J and slope L_0 (right panel). None of the standard

Table 4.7   Constraints on the asymmetry energy (J) and slope (L_0) at saturation density taken from recent literature. The constraints are labeled for convenience.

Reference	J (MeV)	L_0 (MeV)	Designation
[Lattimer and Lim (2013)]	29–32.7	40.5–61.9	J1, L1
[Danielewicz and Lee (2014)]	30.2–33.7	35–70	J2, L2
[Hagen *et al.* (2015)]	25.2–30.4	37.8–47.7	J3, L3

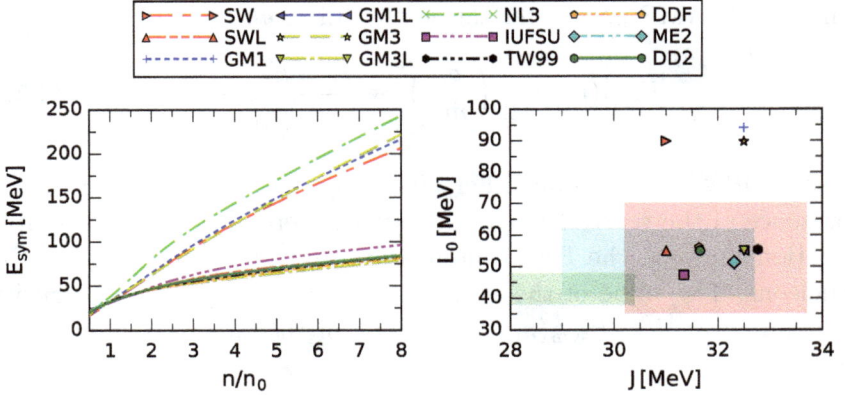

Fig. 4.4   (Left panel) The asymmetry energy E_{sym} as a function of the baryon number density in SNM. (Right panel) The asymmetry energy slope L_0 plotted against the asymmetry energy at saturation density. The cyan shading indicates the constraints J1 and L1, the red shading the constraints J2 and L2, and the green shading the constraints J3 and L3 (see Table 4.7).

RMF parameterizations (excluding IUFSU) are able to satisfy the constraints on L_0 (L1–L3) as expected, though the SW, GM1, and GM3 parameterizations do satisfy the J1 and J2 constraints. The NL3 parameterization has an extremely high $L_0 = 118.5\,\mathrm{MeV}$ and $J = 37.4\,\mathrm{MeV}$ and so does not satisfy L1–L3 or J1–J3 and is outside the bounds of Fig. 4.4. The IUFSU parameterization includes additional nonlinear vector and cross interactions that make it possible to satisfy L1–L3 in addition to J1 and J2. The RMFL and DDRMF parameterizations all satisfy J1–J2 and L1–L2 except TW99 which lies outside of J1. None of the parameterizations studied satisfy J3 or L3 (excluding IUFSU), the recent constraint from Hagen *et al.* [Hagen *et al.* (2015)] derived from *ab initio* calculations of the neutron distribution of ^{48}Ca. Developing a parameterization that satisfies these constraints for studying exotic matter in NSs will be a focus of future work.

The Dirac effective nucleon mass (or effective mass) at saturation density given by

$$m^*(n_0) = m_N - g_{\sigma N}(n_0)\bar{\sigma}(n_0), \qquad (4.55)$$

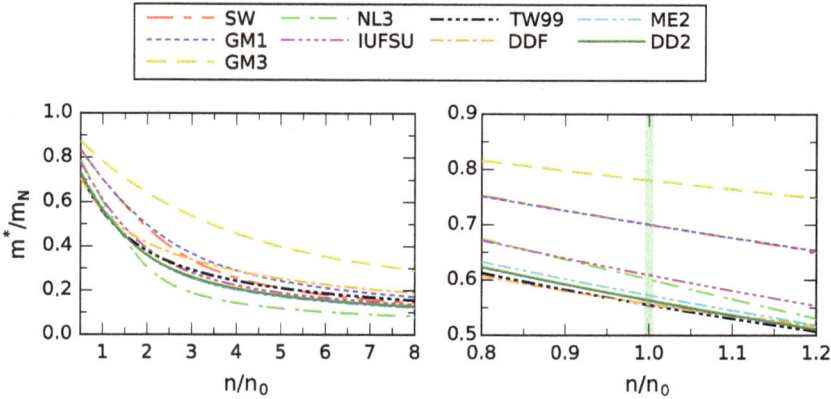

Fig. 4.5 The effective nucleon mass m^*/m_N as a function of baryon number density in SNM. The right panel enlarges the region around the saturation density n_0, and the green shading identifies the saturation density for convenience.

is not well constrained. Values of $\bar{\sigma}$ derived from the scattering of neutrons from lead nuclei suggest a range of $m^*/m_N \approx 0.7$–0.8 [Johnson, Horen, and Mahaux (1987); Mahaux and Sartor (1987); Jaminon and Mahaux (1989)]. However, to reasonably reproduce empirical spin–orbit splittings in finite nuclei a range of $m^*/m_N \approx 0.58$–0.64 was required [Furnstahl, Rusnak, and Serot (1998)]. As a result of this persistent ambiguity the effective mass will not be used as a constraint to eliminate RMF and DDRMF parameterizations from consideration. The effective mass as a function of baryon number density is given in Fig. 4.5 for all parameterizations included in this work. Note the extremely low effective mass of the NL3 parameterization which has a tendency to prohibit models with additional baryonic degrees of freedom from reaching their maximum central densities (see for example [Miyatsu, Cheoun, and Saito (2013)]).

4.3.1.2. *Constraints from heavy-ion collisions*

The EoS of SNM can be constrained directly by experimental data from heavy-ion collisions (HICs) [Klähn *et al.* (2006)]. Danielewicz *et al.* (2002) analyzed experimental elliptic flow data in HICs to deduce bounds on the high density $(2 - 4.6\,n_0)$ EoS of SNM

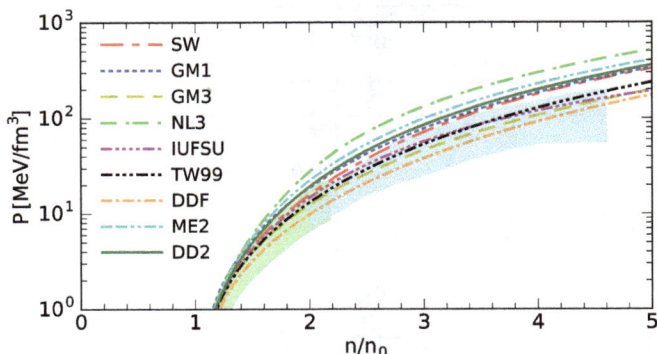

Fig. 4.6   The EoS of SNM as a function of baryon number density in units of the saturation density. The green shaded region indicates the constraint derived from kaon production in heavy-ion collisions [Lynch *et al.* (2009)], and the blue shaded region indicates the constraint derived from the analysis of transverse and elliptical particle flow in heavy-ion collisions [Danielewicz, Lacey, and Lynch (2002)].

represented by the cyan shaded region in Fig. 4.6 [Danielewicz, Lacey, and Lynch (2002)]. These bounds suggest a rather soft EoS at high densities and are typically difficult to satisfy in light of mass constraints to be discussed in the next section. The IUFSU, GM3, TW99, and DDF parameterizations successfully satisfy this constraint, though the DDF parameterization was constructed specifically to do so in [Klahn *et al.* (2006)].

Lynch *et al.* (2009) used kaon production data from HICs to similarly deduce bounds on the $\sim 1 - 2\,n_0$ SNM EoS that are consistent with the low density bounds from the elliptic flow constraint [Lynch *et al.* (2009); Fuchs (2006)]. These bounds are represented in Fig. 4.6 by the green shaded region. As with the elliptic flow constraint, the IUFSU, GM3, TW99, and DDF parameterizations successfully satisfy this constraint, but in addition the SW parameterization created for this work is also consistent with this constraint.

4.3.2. *Constraining the equation of state with neutron star properties*

A viable model for the NS EoS should be capable of reproducing observed NS properties in addition to the properties of bulk nuclear matter at saturation density discussed in the previous sections. For non-rotating NSs at zero temperature the properties of primary importance are the NS mass and radius, which are determined from the EoS by solving the TOV equation (4.40). However, additional constraints from NS properties exist for these conditions. The baryonic mass of a NS has been determined at a precise value of the gravitational mass from observations of a NS in a NS–white dwarf binary system. Also, the population of NSs at masses below $1.5\,M_\odot$ suggests constraints must be placed on the speed of NS cooling, restricting the operation of the direct Urca (DU) enhanced cooling mechanism in low mass NSs. Analysis of this constraint does not require a thermal evolution calculation.

In the following section we will discuss these constraints that emerge from NS observations, and determine if the chosen parameterizations of the NS EoS satisfy them. For now we assume that NS matter consists of neutrons, protons, electrons, and muons (nucleonic or $npe\mu$ matter).

4.3.2.1. *Maximum mass*

The NS EoS is constrained by the requirement that it produces a maximum mass that can account for the highest measured pulsar masses. The significance of this constraint increased tremendously with the precise measurement of two-solar-mass pulsars in 2010 and 2013. In 2010 Demorest *et al.* [Demorest *et al.* (2010)] published the mass of PSR J1614-2230 as $1.97 \pm 0.04\,M_\odot$ at 1σ. However, this value has been recently updated to the lower value $1.928 \pm 0.017\,M_\odot$, also at 1σ, though with a factor of two improvement in the precision [Fonseca *et al.* (2016)]. In 2013 Antoniadis *et al.* [Antoniadis *et al.* (2013)] published the larger mass of PSR J0348+0432 as $2.01 \pm 0.04\,M_\odot$ at 1σ $(1.90 - 2.18\,M_\odot$ at $3\sigma)$. Both of these pulsar masses were precisely measured by exploiting the observed Shapiro delay, a time delay in the radio pulse from the pulsar due to the passage of the signal through the gravitational field of its massive binary companion. These discoveries inevitably led to suggestions that exotic NS cores could be ruled out, as the softer exotic EoS would have difficulty satisfying the high mass constraints.

For convenience, in later sections we will refer to the full $2.01\,M_\odot$ mass of PSR J0348+0432 as the **"strong"** mass constraint, and the $1.90\,M_\odot$, -3σ limit on its mass, the **"weak"** mass constraint. Note that the -1σ limit on the mass of PSR J1614-2230 of $1.911\,M_\odot$ nearly coincides with this weak constraint.

The mass–radius curves for $npe\mu$ EoS with the chosen RMF(L) and DDRMF parameterizations are shown in Fig. 4.7, and the maximum masses are provided in Table 4.8. All of the parameterizations

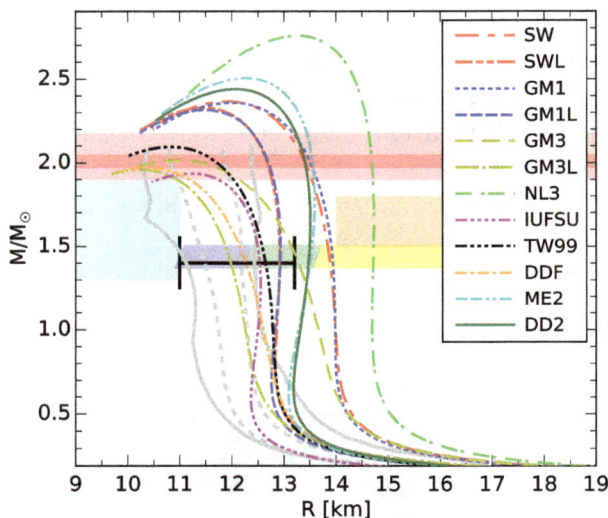

Fig. 4.7   Neutron star mass versus radius for the given RMF, RMFL, and DDRMF parameterizations. The dark (light) red shading indicates the 1σ (3σ) uncertainty range in the mass of PSR J0348+0432 [Antoniadis *et al.* (2013)]. The yellow, green, and purple shading indicate the 1σ, 2σ, and 3σ uncertainty limits respectively on the radius of PSR J0437-4715 [Bogdanov (2013)], with the $1.44 \pm 0.07\,M_\odot$ mass range determined in [Reardon *et al.* (2016)]; the orange shading indicates the radial constraint set by RX J1856.5-3754 [Hambaryan *et al.* (2014)]; the dashed (solid) gray lines represent the 1σ (2σ) limits on the NS radius from Bayesian analyses that included NS observations and properties [Steiner, Lattimer, and Brown (2010)]; the solid black line indicating the radial range from 11–13.2 km was deduced from the confluence of nuclear theory and experimental data [Lattimer and Steiner (2014)]; and finally the cyan shading indicates the approximate constraint region suggested by measurements of three NSs in binaries [Özel, Baym, and Güver (2010)].

Table 4.8   Properties of maximum mass NSs including the maximum mass $M/M_\odot$ in solar mass units, the radius of the maximum mass NS $R_{\max}$, the radius of the $1.4\,M_\odot$ NS $R_{1.4}$, and the maximum baryon number density $n_{\max}$.

Parameterization	$M/M_\odot$	$R_{\max}$ (km)	$R_{1.4}$ (km)	$n_{\max}$ $(1/\mathrm{fm}^3)$
		RMF		
SW	2.37	11.9	13.9	0.87
GM1	2.36	12.0	13.9	0.87
GM3	2.02	10.9	13.3	1.09
NL3	2.76	13.3	14.7	0.67
IUFSU	1.94	11.2	12.5	1.03
		RMFL		
SWL	2.33	11.5	12.9	0.92
GM1L	2.32	11.5	12.9	0.92
GM3L	1.96	10.2	12.0	1.21
		DDRMF		
TW99	2.09	10.8	12.6	1.08
DDF	1.97	10.4	12.3	1.20
DD2	2.44	12.0	13.5	0.84
ME2	2.51	12.3	13.5	0.80

have a maximum mass that exceeds the $\sim 1.93\,M_\odot$ mass of PSR J1614-2230. However, DDF, GM3L, and IUFSU do not fully satisfy the strong mass constraint set by PSR J0348+0432, DDF satisfying the -1σ level with $M_{\max} = 1.97\,M_\odot$, and GM3L and IUFSU satisfying the -3σ level with $M_{\max} = 1.96\,M_\odot$ and $1.94\,M_\odot$ respectively.

Parameterizations were excluded from consideration in this work if it was evident that their associated nucleonic EoS would not satisfy these mass constraints. Therefore, by design at this time the mass constraint does not rule out any of the included parameterizations. However, as previously stated the relevance of this constraint increases significantly with the consideration of exotics such as hyperons, delta isobars, and deconfined quarks.

4.3.2.2. *Radius*

To date there remains a great deal of uncertainty concerning the NS radius. Recent works have combined NS observations with constraints from nuclear experiment and theory to determine bounds on the NS radius. Steiner *et al.* (2010) used Bayesian analysis to combine simultaneous mass–radius observations of six low-mass X-ray binaries (LMXBs) with maximum mass, causality, and spin frequency constraints, providing bounds on the NS mass–radius curve [Steiner, Lattimer, and Brown (2010)]. Bogdanov (2013) used X-ray spectroscopy to determine bounds on the radius of the millisecond pulsar PSR J0437-4715 [Bogdanov (2013)]. Lattimer *et al.* (2014) utilized constraints from nuclear experiment, in particular bounds on E_{sym} and L, to provide limits on the radius of a $1.4 M_\odot$ NS of $R_{1.4} \approx 12.1 \pm 1.1$ km [Lattimer and Steiner (2014); Lattimer (2019)]. These three combined results suggest that $R \gtrsim 11$ km is a reasonable lower limit on the NS radius, but those of Steiner and Lattimer suggest the upper limit may be as low as ~13 km, a constraint that eliminates many stiff RMF EoS. However, the 1σ bounds on PSR J0437-4715 and radial bounds on the isolated NS J1856.6-3754 seem to favor a stiffer EoS [Bogdanov (2013); Hambaryan *et al.* (2014)]. Finally, Ozel *et al.* (2010) analyzed three NSs in double NS binaries and determined that all three have very small radii of 8–11 km, a suggestion at odds with the previous radial determinations presented [Özel, Baym, and Güver (2010)].

Mass–radius curves and the aforementioned radial constraints are shown together in Fig. 4.7. The radius of J0437-4715 really serves only to provide a lower limit on the NS radius, and is satisfied in some way by all parameterizations, but only the extremely stiff NL3 lies in the 1σ range. The radii of SW and GM1 NSs lie outside both the Steiner and Lattimer bounds. However, the addition of the density-dependent isovector coupling $g_{\rho N}(n_0)$ (SWL, GM1L) softens these equations of state reducing their radius at $1.4 M_\odot$ by around 1 km, satisfying the associated constraint. The DD2 and ME2 EoS are slightly too stiff to satisfy the $1.4 M_\odot$ constraint, but successfully satisfy so many other constraints that this alone does not serve to disqualify them. Only the DDF and GM3L EoS truly satisfy the mass–radius bounds from Steiner, but those EoS are soft and will have difficulty satisfying mass constraints with exotic matter cores.

4.3.2.3. *Gravitational and baryonic mass*

Pulsar B of the double pulsar PSR J0737-3039 has a very low and precisely measured mass of $1.249 \pm 0.001 \, M_\odot$, suggesting that it formed in an electron-capture supernova of an O–Ne–Mg white dwarf [Kramer *et al.* (2005)]. Podsiadlowski *et al.* (2005) used numerous core compositions of O–Ne–Mg white dwarfs from the literature to determine the baryonic mass (M_B) of Pulsar B to be between $1.366 - 1.375 \, M_\odot$, assuming little to no mass loss [Podsiadlowski *et al.* (2005)]. M_B was calculated by multiplying the total baryon number N_B, given by (4.42), by the atomic mass unit (931.50 MeV). Requiring that these two measurements be simultaneously satisfied very tightly constrains the NS EoS. However, a later work by Kitaura *et al.* (2006) simulated electron-capture supernovae of O–Ne–Mg cores and determined that mass loss was likely, yielding a baryonic mass of between $1.358 - 1.362 \, M_\odot$ for Pulsar B, slightly weakening the constraint [Kitaura, Janka, and Hillebrandt (2006)].

The gravitational mass is plotted against the baryonic mass for all parameterizations in Fig. 4.8. The DD parameterizations are all

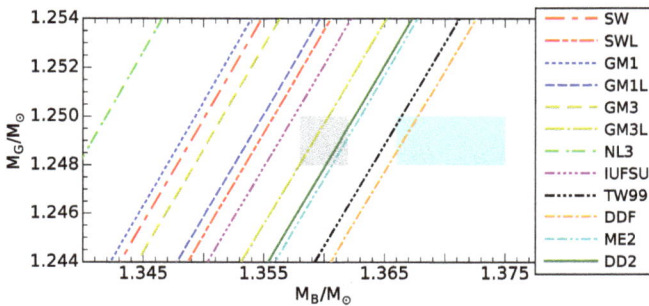

Fig. 4.8   Neutron star gravitational mass versus total baryonic mass, a relationship constrained by observations of the double pulsar PSR J0737-3039. The gray and cyan shaded rectangles represent the constraints from [Kitaura, Janka, and Hillebrandt (2006)] and [Podsiadlowski *et al.* (2005)] respectively.

more consistent with this constraint than those of RMF(L). Only the DDF and TW99 EoS even partially satisfy the tighter constraint of Podsiadlowski *et al.* (2005), but the DD2 and ME2 parameterizations are at least consistent with the mass-loss constraint from Kitaura *et al.* (2006). The SW, GM1, GM3, and NL3 standard RMF EoS are extremely inconsistent with this constraint, but significantly greater consistency is seen with their RMFL counterparts.

4.3.2.4. *The direct Urca process*

The direct Urca (DU) process is the most efficient cooling process in NSs at temperatures below $T \approx 10^{12}$ K [Lattimer *et al.* (1991)]. If the DU process operates in the core of a NS it will cool it beyond the level of thermal detection within a few years. Evidence suggests that NSs with masses below $M \lesssim 1.5\,M_\odot$ should remain visible for longer than the DU process would allow, leading to the constraint that the DU process be potentially active only in NSs with $M \gtrsim 1.5\,M_\odot$ (we will focus on this "strong" constraint, though $M \gtrsim 1.35\,M_\odot$ has been proposed as a "weak" constraint) [Klahn *et al.* (2006); Blaschke, Grigorian, and Voskresensky (2004); Popov *et al.* (2006)].

The nucleonic direct Urca (NDU) processes are those of β decay and inverse β decay,

$$n \to p + e^- + \bar{\nu}_e, \quad p + e^- \to n + \nu_e. \tag{4.56}$$

For these processes to proceed momentum conservation must be satisfied, regulated by the triangle inequality with cyclic permutations over ijk,

$$k_{F_i} + k_{F_j} \geq k_{F_k}. \tag{4.57}$$

In a NS neutrons typically exist in much greater number and with a much higher Fermi momentum than protons or electrons, so the relevant triangle inequality is generally

$$k_{F_p} + k_{F_e} \geq k_{F_n}. \tag{4.58}$$

The DU threshold is commonly approximated in simple terms of the electron fraction, but this approximation does not account for the presence of charged hyperons or Δ baryons, so instead the threshold

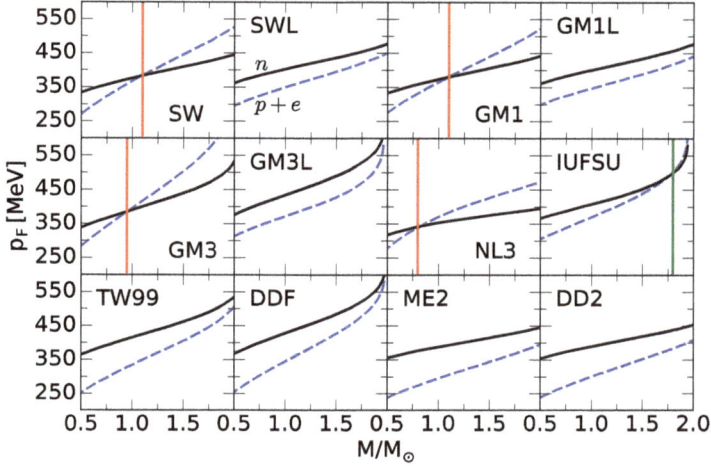

Fig. 4.9 Fermi momentum of the neutron (solid black line) and the combined Fermi momentum of the proton and electron (dashed blue line). The NDU process operates when the dashed blue line crosses above the solid black line (see (4.58)).

for DU onset will be determined by direct evaluation of the triangle inequalities for consistency.

Figure 4.9 shows the Fermi momenta for the terms in (4.58) for all parameterizations. All four standard RMF parameterizations (SW, GM1, GM3, NL3) violate the DU constraint by a significant margin, with NDU onset at masses below $1.1\,M_\odot$. The NDU process does not occur at all in any of the RMFL and DDRMF parameterizations, but occurs in the IUFSU parameterization at a constraint satisfying $1.8\,M_\odot$.

4.3.3. *Summary: Constraining the equation of state*

In this section we have introduced a number of constraints on the NS EoS from both the study of nuclear matter and the properties of NSs deduced from observations. These constraints and the ability or inability of the chosen parameterizations to satisfy them are summarized in Tables 4.9 and 4.10. No parameterization satisfies all of the constraints, but this is technically impossible since they are not all

Table 4.9   Constraints on the EoS of SNM. A "+" represents satisfaction of the given constraint, while a "−" indicates the constraint was not satisfied. The bounds on K_0, J, and L_0 identified by K#, J#, and L# can be found in Tables 4.6 and 4.7. SNM EoS refers to the constraints deduced from kaon production (K^+) and elliptic flow (Flow) in HICs. Refs.: I [Shlomo, Kolomietz, and Colo (2006)], II [Stone, Stone, and Moszkowski (2014)], III [Lattimer and Lim (2013)], IV [Danielewicz and Lee (2014)], V [Hagen *et al.* (2015)], VI [Lattimer and Lim (2013)], VII [Danielewicz and Lee (2014)],VIII [Hagen *et al.* (2015)], IX [Lynch *et al.* (2009)], X [Danielewicz, Lacey, and Lynch (2002)].

| Parameterization | K_0 | | J | | | L_0 | | | SNM EoS | |
	K1	K2	J1	J2	J3	L1	L2	L3	K^+	Flow
					RMF					
SW	+	+	+	+	−	−	−	−	+	−
GM1	−	+	+	+	−	−	−	−	−	−
GM3	+	−	+	+	−	−	−	−	+	+
NL3	−	+	−	−	−	−	−	−	−	−
IUFSU	+	−	+	+	−	+	+	+	+	+
					RMFL					
SWL	+	+	+	+	−	+	+	−	+	−
GM1L	−	+	+	+	−	+	+	−	−	−
GM3L	+	−	+	+	−	+	+	−	+	+
					DDRMF					
TW99	+	−	+	+	−	+	+	−	+	+
DDF	+	−	−	+	−	+	+	−	+	+
DD2	+	−	+	+	−	+	+	−	−	−
ME2	+	+	+	+	−	+	+	−	−	−
Ref.	I	II	III	IV	V	VI	VII	VIII	IX	X

necessarily consistent with one another. This is particularly evident in the distribution of radial constraints shown in Fig. 4.7.

One of the primary motivations for this section was to illustrate the fact that the standard RMF approach and its associated parameterizations are clearly inadequate for the study of exotic matter in NSs. The standard RMF parameterizations (excluding GM3) yield

Table 4.10 Constraints deduced from NS observations. A "+" represents satisfaction of the given constraint, while a "−" indicates the constraint was not satisfied. "M_G vs. M_B" refers to the baryonic mass constraint imposed from observations of PSR J0737-3039 by Podsiadlowski *et al.* (Pod.) and Kitaura *et al.* (Kit.). $R_{1.4}$, R^*, and R^{**} refer to the radial constraints on a 1.4 $M_\odot$ NS from nuclear systematics, from observations of PSR J0437-4715, and from observations of NS J1856.6-3754, respectively. NDU refers to the requirement that the DU process occur only in NSs with $M > 1.5\,M_\odot$. Refs.: I [Antoniadis *et al.* (2013)], II [Podsiadlowski *et al.* (2005)], III [Kitaura, Janka, and Hillebrandt (2006)], IV [Lattimer and Lim (2013)], V [Bogdanov (2013)], VI [Hambaryan *et al.* (2014)], VII [Klähn *et al.* (2006)].

Parameterization	Mass of PSR J0348+0432					M_G vs. M_B		$R_{1.4}$	R^*			R^{**}	NDU
	-3σ	-1σ	2.01	$+1\sigma$	$+3\sigma$	Pod.	Kit.		1σ	2σ	3σ		
RMF													
SW	+	+	+	+	+	−	−	−	−	+	−	−	−
GM1	+	+	+	+	+	−	−	−	+	+	−	−	−
GM3	+	+	+	−	−	−	−	−	+	+	+	−	−
NL3	+	+	+	+	+	−	−	−	+	−	−	+	−
IUFSU	+	−	−	−	−	−	−	+	+	+	+	+	+
RMFL													
SWL	+	+	+	+	+	−	−	+	−	+	+	−	+
GM1L	+	+	+	+	+	−	−	+	−	+	+	−	+
GM3L	+	−	−	−	−	−	+	+	−	−	+	−	+
DDRMF													
TW99	+	+	+	+	−	+	−	+	−	−	+	−	+
DDF	+	+	−	−	−	+	−	+	−	−	+	−	+
DD2	+	+	+	+	+	−	+	−	−	+	−	−	+
ME2	+	+	+	+	+	−	+	−	−	+	−	−	+
Ref.	I					II	III	IV	V			VI	VII

a very stiff EoS that can produce massive NSs capable of satisfying even the highest mass constraint ($M_{\text{PSR J0348+0432}} = 2.18\,M_\odot$ at upper $+3\sigma$ boundary). However, these parameterizations are not consistent with the increasingly precise bounds on the slope of the asymmetry energy at saturation density (L_0), they produce large radii inconsistent with the bounds on $R_{1.4}$, they violate the DU constraint by a significant margin, and they are drastically inconsistent with the baryonic mass constraint. In addition, the NL3 parameterization is extremely stiff and has an outlying saturation asymmetry energy (J) that is terribly inconsistent with constraints.

Augmenting the standard RMF approach with a density-dependent isovector meson–baryon coupling constant (RMFL) has a substantial effect on the satisfaction of constraints. In contrast to standard RMF, the RMFL parameterizations successfully satisfy the constraints on L_0, they produce mass–radius relations that satisfy the $R_{1.4}$ constraint, and the NDU process does not operate at any mass satisfying the DU constraint. Therefore the standard RMF parameterizations (excluding NL3) will be replaced by their RMFL counterparts and utilized in the analyses to follow.

The fully density-dependent DDRMF parameterizations are fairly stiff so that they easily satisfy the mass constraints but yield radii that are slightly inconsistent with the $R_{1.4}$ constraint. They are very consistent with the saturation properties of SNM and as with RMFL the NDU process does not proceed at any mass satisfying the DU constraint. In addition, the DDRMF parameterizations almost exclusively satisfy the baryonic mass constraint (exception is GM3L). These observations coupled with the fact that these models are in addition parameterized to reproduce properties of finite nuclei suggests that they are well suited for the further study of exotic matter in NSs.

Finally, of primary importance in the study of exotic matter will be the mass constraint, as the softening of the EoS due to the introduction of additional baryonic degrees of freedom makes the satisfaction of this constraint more difficult. It will be shown in the next section that the parameterizations consistent with the elliptic flow HIC constraint are not stiff enough to simultaneously satisfy the mass constraint when exotic matter is considered. This observation, coupled with the fact that the elliptic flow constraint was deduced with calculations that are EoS-dependent lead us to regard this constraint as being of questionable utility [Klahn *et al.* (2006); Danielewicz, Lacey, and Lynch (2002)].

4.4. Hyperons in Neutron Star Matter

Cold nucleonic NS matter is in the ground state and any increase in density must coincide with an increase in the nucleon Fermi momentum due to the Pauli Exclusion Principle. As the nucleon chemical potential increases it will eventually become more energy efficient to populate hyperonic degrees of freedom, slowing the increase of both the nucleon and electron Fermi momenta, the charge neutrality condition being satisfied by the newly created negatively charged hyperons rather than high energy leptons. This hyperonization of NS matter typically results in a considerable softening of the EoS compared to a purely nucleonic EoS, leading to an inevitable decrease in the NS maximum mass. However, a hyperonic EoS is subject to all the same constraints as described in Sec. 4.3 for purely nucleonic EoS, including the 2.01 $M_\odot$ maximum mass constraint.

In order to study the hyperonization of the NS core we will utilize the nuclear parameterizations that most successfully satisfied the constraints discussed in Sec. 4.3. Hyperons do not typically appear until at least twice the nuclear saturation density, so the satisfaction of saturation constraints will not need to be re-examined. Constraints on macroscopic NS properties such as the maximum mass and the lack of enhanced cooling in low mass NSs must still be satisfied. These constraints, as well as the available experimental data on the hyperons will be utilized to constrain the meson–hyperon coupling constants. The degree of hyperonization will be determined for different parameterizations, and its dependence on certain nuclear saturation properties will be analyzed.

The properties of the hyperons in the spin-$\frac{1}{2}$ baryon octet are summarized in Table 4.11. The hyperons are distinct from nucleons in that they all consist of at least one strange quark, and the 95 MeV

Table 4.11   Properties of the hyperons in the spin-$\frac{1}{2}$ baryon octet. Given are the mass m, electric charge q, the total spin J, the z component of the isospin I_3, the constituent quarks, and the strangeness S. The masses are those published by the Particle Data Group [Patrignani *et al.* (2016)].

Hyperon	m (MeV)	q (e)	J	I_3	Quarks	S
Λ	1115.683	0	1/2	0	uds	-1
Σ^+	1189.37	1	1/2	1	uus	-1
Σ^0	1192.642	0	1/2	0	uds	-1
Σ^-	1197.449	-1	1/2	-1	dds	-1
Ξ^0	1314.83	0	1/2	1/2	uss	-2
Ξ^-	1321.31	-1	1/2	$-1/2$	dss	-2

mass of the strange quark is more than 40 times that of the up quark ($m_u \approx 2.2$ MeV) and 20 times that of the down quark ($m_d \approx 4.7$ MeV).

4.4.1.   *Meson–hyperon coupling constants*

In this section we will discuss the determination of the meson–hyperon coupling constants g_{iY}, for $i \in \{\sigma, \omega, \rho\}$ and $Y \in \{\Lambda, \Sigma^+, \Sigma^0, \Sigma^-, \Xi^0, \Xi^-\}$. Since the value of any g_{iY} is usually determined in terms of meson–nucleon couplings g_{iN} of the chosen parameterization, the preferred representation of the hyperon couplings is

$$x_{iY} = g_{iY}/g_{iN}. \tag{4.59}$$

The meson–hyperon couplings are not well constrained experimentally compared to those of the nucleons. However, the scalar meson–hyperon couplings ($x_{\sigma Y}$) can be constrained by the available experimental data on hypernuclei, but their calculation first requires the determination of the vector meson–hyperon couplings ($x_{\omega Y}$).

4.4.1.1.   *Vector meson–hyperon coupling constants*

It has been shown that in SU(3) flavor symmetry the vector meson–hyperon coupling constants can be expressed in terms of three parameters: the ratio of the meson singlet and octet couplings to the baryon octet $z = g_8/g_1$, the vector mixing angle θ_V, and the symmetric/antisymmetric vector coupling ratio α_V [Weissenborn, Chatterjee, and Schaffner-Bielich (2012a)]. These expressions are as follows [Dover and Gal (1984)]:

$$x_{\omega\Lambda} = \frac{1 - \frac{2}{\sqrt{3}} z \tan \theta_V (1 - \alpha_V)}{1 - \frac{1}{\sqrt{3}} z \tan \theta_V (1 - 4\alpha_V)}, \tag{4.60}$$

$$x_{\omega\Sigma} = \frac{1 + \frac{2}{\sqrt{3}} z \tan \theta_V (1 - \alpha_V)}{1 - \frac{1}{\sqrt{3}} z \tan \theta_V (1 - 4\alpha_V)}, \tag{4.61}$$

$$x_{\omega\Xi} = \frac{1 - \frac{1}{\sqrt{3}} z \tan \theta_V (1 + 2\alpha_V)}{1 - \frac{1}{\sqrt{3}} z \tan \theta_V (1 - 4\alpha_V)}. \tag{4.62}$$

Under the assumption that the $S = -1$ hyperons (Λ and Σ) both couple equally to the vector meson, $\alpha_V = 1$, and the coupling relations between the vector meson–hyperon couplings can be simplified to the following [Weissenborn, Chatterjee, and Schaffner-Bielich (2012a); Miyatsu, Cheoun, and Saito (2013); Dover and Gal (1984); Oertel *et al.* (2015); Schaffner *et al.* (1994)]:

$$x_{\omega\Lambda} = x_{\omega\Sigma} = \frac{x_{\omega\Xi}}{1 - \sqrt{3}\, z \tan(\theta_V)} = \frac{1}{1 + \sqrt{3}\, z \tan(\theta_V)}. \tag{4.63}$$

The authors of [Rijken, Nagels, and Yamamoto (2010)] used a broken SU(3) flavor symmetry to describe NN, YN, and YY interactions. This sophisticated interaction model, known as the Nijmegen extended-soft-core (ESC08) model, was used to fit available YN scattering data and extract the mixing angle $\theta_V = 37.50°$ and coupling ratio $z = 0.1949$. Inserting these values into (4.63) yields the ESC08 vector meson–hyperon coupling constants listed in Table 4.12.

In SU(6) spin–flavor symmetry, a special case of SU(3) also known as the quark counting model, the mixing angle $\theta_{V,\mathrm{ideal}} \approx 35.26°$ is chosen so that the vector ω meson is a pure state of up and down quarks. Requiring that the strange vector meson ϕ (not included in this work) not couple to the nucleons necessarily results in the coupling ratio $z = 1/\sqrt{6} \approx 0.40825$. Applying these SU(3) parameters,

Table 4.12   The vector mixing angle θ_V, vector coupling ratio α_V, meson singlet to octet coupling ratio z, and vector meson–hyperon coupling constants given by the SU(6) symmetry and the SU(3) ESC08 model [Rijken, Nagels, and Yamamoto (2010)].

Symmetry	θ_V	α_V	z	$x_{\omega\Lambda} = x_{\omega\Sigma}$	$x_{\omega\Xi}$
SU(6)	35.26°	1	0.40825	2/3	1/3
ESC08	37.50°	1	0.1949	0.79426	0.58852

the strength of the vector meson–hyperon couplings compared to that of the nucleon is reduced by a third per constituent strange quark,

$$x_{\omega Y} = 1 + \frac{S_Y}{3}. \tag{4.64}$$

These SU(6) vector meson–hyperon coupling constants are listed in Table 4.12.

4.4.1.2. *Scalar meson–hyperon coupling constants*

Given the vector meson–hyperon couplings the scalar meson–hyperon coupling constants $(x_{\sigma Y})$ can be fit to the hyperon potential depths determined from experimental data on hypernuclei, nuclei that have at least one nucleon replaced by a hyperon. Hypernuclei are typically created in the laboratory by bombarding certain atoms with mesons (K,π) or electrons [Chatterjee and Vidana (2016)]. The expression for the hyperon potential at nuclear saturation density in SNM is given by

$$U_Y^{(N)}(n_0) = x_{\omega Y} g_{\omega N}\bar\omega - x_{\sigma Y} g_{\sigma N}\bar\sigma + \Sigma_r, \tag{4.65}$$

and can be solved for $x_{\sigma Y}$ given $x_{\omega Y}$. For standard RMF parameterizations there is no contribution from the rearrangement energy, $\Sigma_r = 0$.

The accepted value of the Λ potential, $U_\Lambda^{(N)} \approx -28\,\text{MeV}$, has been deduced from numerous experimental results of Λ binding energies in hypernuclei (see [Chatterjee and Vidana (2016)] and references therein). There are far fewer results from Ξ hypernuclei, but those that exist suggest an attractive potential $U_\Xi^{(N)} \approx -18\,\text{MeV}$ [Miyatsu, Cheoun, and Saito (2013); Oertel *et al.* (2015); Gomes *et al.* (2015); Lopes and Menezes (2014); Tolos, Centelles, and Ramos (2017)]. Finally, since to date the existence of Σ hypernuclei has not been confirmed, the Σ potential is generally assumed to be repulsive and taken to be $U_\Sigma^{(N)} \approx +30\,\text{MeV}$ [Miyatsu, Cheoun, and Saito (2013); Gomes *et al.* (2015); Lopes and Menezes (2014); Tolos, Centelles, and Ramos (2017); Maslov, Kolomeitsev, and Voskresensky (2016)]. However, it will be shown in Sec. 4.4.4 that if the Σ potential is indeed repulsive its strength likely has little effect on NS properties. The SU(6) and ESC08 scalar meson–hyperon coupling constants used in this work are listed in Table 4.13.

4.4.1.3. *Isovector meson–hyperon coupling constants*

The relative isovector meson–hyperon coupling constants are typically scaled by the hyperon isospin as follows [Weissenborn, Chatterjee, and Schaffner-Bielich (2012a); Miyatsu, Cheoun, and Saito (2013); Maslov, Kolomeitsev, and Voskresensky (2016)],

$$x_{\rho Y} = 2|I_{3Y}|. \tag{4.66}$$

Symmetry arguments similar to those used to determine the vector meson–hyperon couplings have been found to result in isovector

Table 4.13 Scalar meson–hyperon coupling constants given by (4.65) with the vector meson–hyperon coupling constants given by the SU(6) symmetry and the ESC08 model in Table 4.12.

Parameterization	SU(6)			ESC08		
	$x_{\sigma\Lambda}$	$x_{\sigma\Sigma}$	$x_{\sigma\Xi}$	$x_{\sigma\Lambda}$	$x_{\sigma\Sigma}$	$x_{\sigma\Xi}$
SWL	0.6131	0.4072	0.3208	0.7115	0.5773	0.5276
GM1L	0.6110	0.4051	0.3197	0.7089	0.5030	0.5155
GM3L	0.6060	0.3253	0.3224	0.6961	0.4153	0.5025
IUFSU	0.6108	0.4530	0.3163	0.7131	0.5553	0.5209
TW99	0.6123	0.4735	0.3182	0.7157	0.5769	0.5250
DDF	0.6153	0.4762	0.3200	0.7191	0.5800	0.5277
DD2	0.6152	0.4740	0.3210	0.7185	0.5773	0.5276
ME2	0.6151	0.4708	0.3225	0.7175	0.5733	0.5274

meson–hyperon couplings that lead to inconsistencies with asymmetry energy constraints, and are thus not employed [Oertel *et al.* (2015)].

4.4.2. *The hyperonic equation of state*

Inclusion of the hyperonic degrees of freedom reduces the nucleon Fermi momenta and softens the equation of state to varying degree depending on the meson–hyperon coupling constants and the chosen nuclear parameterization. In Figs. 4.10 and 4.11 the EoS of RMF(L) and DDRMF parameterizations are shown for nucleonic matter and hyperonic matter in SU(6) symmetry and in the SU(3) ESC08 model. The softening of the hyperonic EoS compared to the nucleonic EoS is dramatic in both SU(6) and SU(3), however the larger ESC08 vector couplings (see Table 4.12) lead to a significantly stiffer EoS for all parameterizations. The hyperonic equation of state is constrained indirectly through the need to satisfy the maximum mass constraint discussed in Sec. 4.3.2.1.

The TOV equation is solved for the hyperonic EoS in Figs. 4.10 and 4.11 to determine the mass–radius relations shown in Fig. 4.12. Only the stiffest DDRMF parameterization (ME2) is able to satisfy the mass constraint from PSR J0348+0432 at the 3σ level if the hyperons couple to the vector meson in SU(6) symmetry. If instead the ESC08 vector meson–hyperon couplings are chosen only half of the parameterizations are able to satisfy the mass constraint (at the

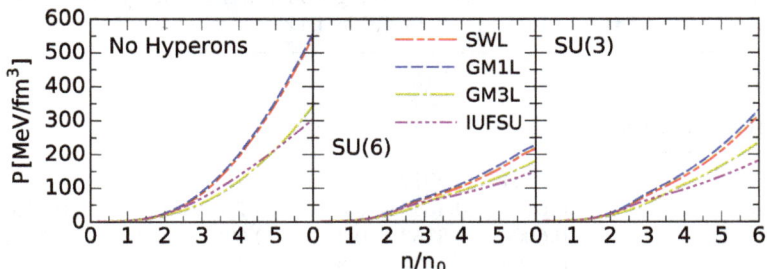

Fig. 4.10 Equations of state for RMFL and NLRMF parameterizations with nucleons only (*left*), nucleons and hyperons in SU(6) symmetry (*middle*), and nucleons and hyperons in the ESC08 model (*right*).

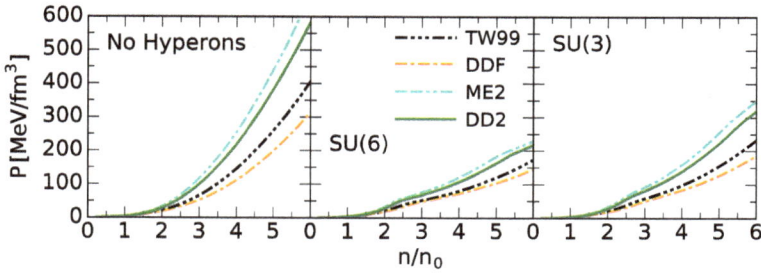

Fig. 4.11 Equations of state for DDRMF parameterizations with nucleons only (*left*), nucleons and hyperons in SU(6) symmetry (*middle*), and nucleons and hyperons in the ESC08 model (*right*).

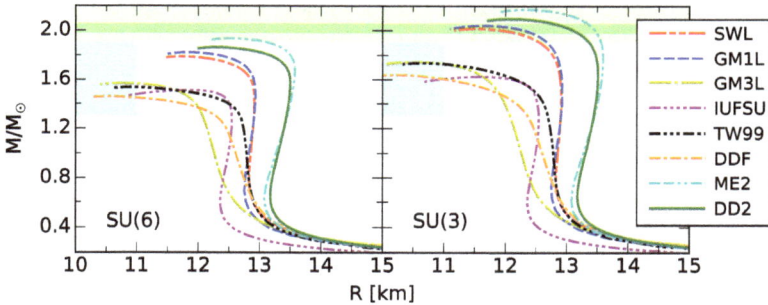

Fig. 4.12 Neutron star mass versus radius for the given RMFL, NLRMF, and DDRMF parameterizations. The vector meson–hyperon coupling constants are computed in SU(6) symmetry (*left*) and in the ESC08 model (*right*). The dark (light) green shading indicates the 1σ (3σ) uncertainty range in the mass of PSR J0348+0432. The cyan shading indicates the approximate constraint region suggested by measurements of three NSs in binaries compiled in Ozel *et al.* (2010) [Özel, Baym, and Güver (2010)].

1σ level), while the other half fall at least $0.15\,M_\odot$ short of the 3σ level. The DD2 and ME2 parameterizations easily satisfy the mass constraint, both with maximum masses above the $2.05\,M_\odot$ upper limit of the 1σ range. The SWL parameterization itself was fixed to have a maximum mass of exactly $2.01\,M_\odot$ with ESC08 meson–hyperon vector couplings, and the GM1L parameterization maximum mass falls slightly above that of SWL. The properties of the maximum mass NSs are summarized in Table 4.14.

Table 4.14   Properties of maximum mass NSs with vector meson–hyperon coupling constants in SU(6) symmetry and in the ESC08 model (Fig. 4.12). Properties include the maximum mass $M/M_\odot$ in solar mass units, the radius R in km, and the maximum baryon number density $n_b^{\max}$ in $1/\mathrm{fm}^3$.

Parameterization	SU(6)			ESC08		
	$M/M_\odot$	R	$n_b^{\max}$	$M/M_\odot$	R	$n_b^{\max}$
SWL	1.78	11.7	0.96	2.01	11.5	0.98
GM1L	1.82	11.8	0.93	2.04	11.6	0.95
GM3L	1.56	10.8	1.18	1.74	10.6	1.20
IUFSU	1.51	11.8	0.87	1.62	11.6	0.94
TW99	1.54	10.9	1.19	1.74	10.7	1.22
DDF	1.45	10.5	1.30	1.63	10.2	1.36
DD2	1.86	12.3	0.88	2.09	12.1	0.89
ME2	1.93	12.5	0.83	2.17	12.3	0.85

The failure of the GM3L, IUFSU, TW99, and DDF parameterizations to satisfy the mass constraint with vector meson–hyperon couplings in the stiffer SU(3) symmetry serves to eliminate them as plausible hyperonic EoS, and they will be disregarded in further analyses of hyperonic NS matter. Further, the ME2 and DD2 parameterizations have very similar saturation properties, so the ME2 parameterization may be disregarded in some analyses in favor of the DD2 parameterization as it has a lower maximum mass and so may serve as the more conservative choice.

Moving forward with the SWL, GM1L, and DD2 parameterizations, one can examine the relative particle number densities to locate the onset of hyperonization and further analyze the effects on the EoS. The relative number density of the hyperons with SU(6) vector couplings is shown in Fig. 4.13. The Λ is the first hyperon to appear at around $2.3 - 2.6\, n_0$ for all parameterizations, followed shortly by the Ξ^- in all cases, and this onset can be seen in the immediate and substantial softening of the EoS at those onset densities in Figs. 4.10 and 4.11. The Λ is favored as it is charge and isospin neutral and immediately slows the increase in the prominent neutron Fermi momentum, decreasing its relative number density. The Ξ^- follows,

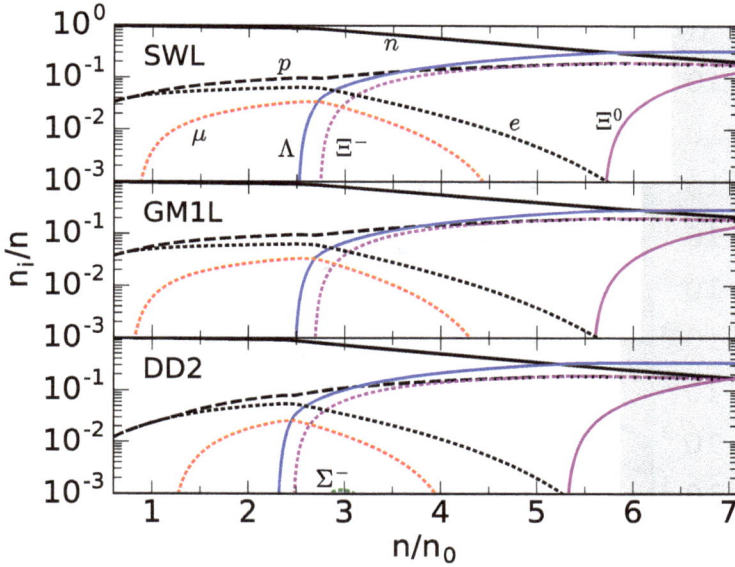

Fig. 4.13   The relative number density of particles as a function of baryon number density (in units of the saturation density). Included are nucleons, leptons, and hyperons with vector meson–hyperon couplings given in SU(6) symmetry. The gray shading indicates baryon number densities beyond the maximum mass star for the given parameterization.

reducing the overall energy of the system by allowing charge neutrality to be satisfied to a greater degree among baryons, lowering the electron population and its associated Fermi momentum. The Σ^- appears only briefly, but the Σ hyperons are in general disfavored due to their repulsive saturation potential. Eventually charge neutrality is accomplished primarily by the proton and Ξ^-, with electrons present in vanishing quantity. Finally, as the Fermi momenta of the Λ and the neutron grow exceedingly large the massive Ξ^0 is energetically favored and appears at around $5 - 5.5 \, n_0$, the resultant softening being particularly visible for the DD2 (and ME2) EoS in Fig. 4.11.

Figure 4.14 shows the relative number density of the hyperons with the ESC08 vector couplings. The increase in the repulsive baryon–hyperon interaction through the vector couplings delays

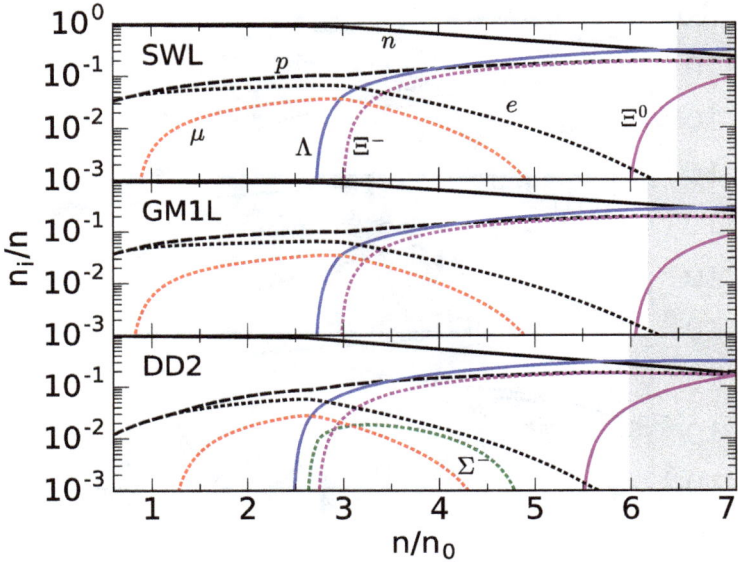

Fig. 4.14   The relative number density of particles as a function of baryon number density (in units of the saturation density). Included are nucleons, leptons, and hyperons with vector meson–hyperon couplings given by the ESC08 model. The gray shading indicates baryon number densities beyond the maximum mass star for the given parameterization.

the onset of hyperonization in each case by about $0.2\,n_0$ compared to SU(6), stiffening the low density equation of state. The order of hyperon onset mimics that seen in SU(6) symmetry, with the exception of the Σ^- which precedes the Ξ^- in DD2 but once again does not appear in substantial numbers compared to other hyperons before vanishing at around $5\,n_0$. The maximum baryon number density achieved is also higher by $0.1 - 0.2\,n_0$ compared to SU(6), further contributing to an increase in maximum mass.

The absence of the Σ^- hyperon in the SWL and GM1L parameterizations is mainly due to the greater magnitude of the isovector field compared to DD2 (see (4.19) and Fig. 4.15). This increases the chemical potential of the Σ^- compared to that of the Ξ^- due to the $\Gamma_{\rho B} I_{3B} \rho$ term (4.35), the Σ^- having twice the negative isospin of the Ξ^- effectively doubling the isovector contribution. If future

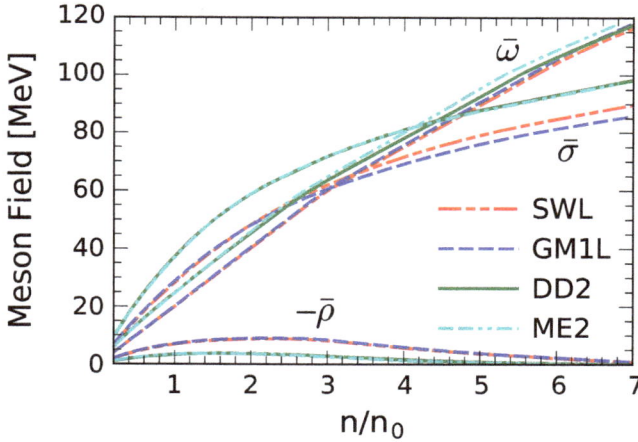

Fig. 4.15   The meson mean fields as a function of baryon number density (in units of the saturation density). Hyperons are included in the ESC08 model.

experiments find evidence of either a less repulsive or an attractive Σ potential the Σ^- will compete with or replace the Ξ^- altogether in the NS core. The effect hypernuclear potentials have on NS properties will be examined further in Sec. 4.4.4.

In order to quantify the degree of hyperonization in the NS core we calculate the strangeness fraction f_S, the ratio of strange quarks to the total number of quarks in a NS given by

$$f_S = \frac{N_s}{N_u + N_d + N_s} = \frac{N_s}{3N_B}. \tag{4.67}$$

The strangeness fraction must be calculated concurrently with the solution of the TOV equation (4.40) by integrating the following,

$$\frac{dN_s}{dr} = \frac{4\pi r^2}{\sqrt{1 - 2m(r)/r}} \sum_B n_B(r)|S_B|, \tag{4.68}$$

where $m(r)$ is given by integration of (4.41), S_B is the strangeness of baryon species B, $G = c = 1$, and N_B is the total number of baryons in the NS given by (4.42). The strangeness fractions of the maximum mass NSs with the vector meson–hyperon coupling constants given by the SU(6) symmetry and SU(3) ESC08 model are given in Table 4.15.

Table 4.15   Maximum strangeness fraction
(as a percentage) of the parameterizations
in Fig. 4.21 with the vector meson–hyperon
couplings given by the SU(6) symmetry and
the SU(3) ESC08 model.

Parameterization	$f_S^{\mathrm{SU}(6)}$ (%)	f_S^{ESC08} (%)
SWL	5.39	5.68
GM1L	4.84	4.78
DD2	5.50	5.86
ME2	4.99	5.36

The SU(3) ESC08 model produces larger strangeness fractions than
SU(6) for the SWL, DD2, and ME2 parameterizations due to the
slightly higher baryon number densities reached in the core. The
DD2 parameterization achieves the highest strangeness fraction at
$f_S^{\mathrm{max}} = 5.86\%$.

Finally, it should be verified that the presence of hyperons does
not cause too early an onset of enhanced cooling due to DU processes for any of the remaining parameterizations. The relevant Fermi
momenta are shown as a function of mass in Fig. 4.16, and it is clear
that the triangle inequality for the nucleonic DU process is not satisfied at less than the maximum mass NS for all parameterizations.
However, similar β decay processes can occur with hyperons, and
those relevant to the particle (hyperonic) composition of the remaining parameterizations are summarized in Table 4.16. Though the
neutrino emissivities due to hyperonic DU are substantially smaller
than that of neutron β decay, they are still much larger than those
of the modified DU processes, on the order $\sim 10^6 T_9^2$ [Prakash *et al.*
(1992)]. Figure 4.16 shows that the triangle inequalities related to
the hyperonic DU processes in Table 4.16 are always satisfied for
all three parameterizations, meaning they proceed as soon as the
hyperons are populated. Should the strong DU constraint be applied
equally to hyperonic processes, it is still satisfied but only just so, as
the Λ appears at or slightly above the $1.5 \, M_\odot$ limit.

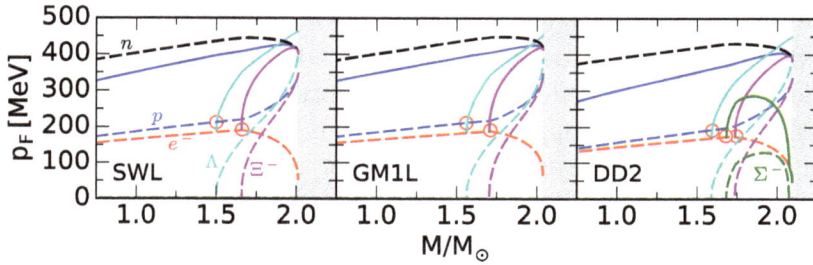

Fig. 4.16   Fermi momentum of baryons as a function of NS mass with vector meson–hyperon couplings given by the ESC08 model. The dashed lines indicate the Fermi momenta of the individual labeled baryons, while solid lines indicate the addition of the electron Fermi momenta to that of the baryon with the same label color. Red circles indicate the onset of a DU process. The gray shading indicates the region beyond the maximum mass NS.

Table 4.16   The hyperon direct Urca processes, typical limiting triangle inequalities, and neutrino emissivities relative to neutron β decay [Prakash *et al.* (1992)].

Direct Urca Process	Triangle Inequality	R
$n \to p + e^- + \bar{\nu}$	$p_{F_p} + p_{F_e} \geq p_{F_n}$	1
$\Lambda \to p + e^- + \bar{\nu}$	$p_{F_\Lambda} + p_{F_e} \geq p_{F_p}$	0.0394
$\Sigma^- \to \Lambda + e^- + \bar{\nu}$	$p_{F_{\Sigma^-}} + p_{F_e} \geq p_{F_\Lambda}$	0.2055
$\Xi^- \to \Lambda + e^- + \bar{\nu}$	$p_{F_{\Xi^-}} + p_{F_e} \geq p_{F_\Lambda}$	0.0175

4.4.3.   *The vector meson–hyperon coupling space*

In the previous section we showed that in order to satisfy the strong mass constraint it is necessary to increase the values of the vector meson–hyperon coupling constants beyond those determined in SU(6) symmetry. The larger vector couplings provided by the ESC08 model in SU(3) symmetry were shown to be sufficient for the satisfaction of the strong mass constraint in the SWL, GM1L, DD2, and ME2 parameterizations. However, these are just two discrete possibilities and provide little information regarding the vector meson–hyperon coupling space as a whole. We would like to see how the mass of PSR

J0348+0432 constrains the vector meson–hyperon couplings beyond the SU(6) and ESC08 determinations. In addition, we are interested in how the onset and degree of hyperonization depend on the vector meson–hyperon couplings. To this end we will systematically explore the parameter space of the vector meson–hyperon coupling constants. Meaningfully investigating the entire three dimensional coupling space of $x_{\omega\Lambda}$, $x_{\omega\Sigma}$, and $x_{\omega\Xi}$ is a bit unrealistic and not theoretically justified, so instead we will explore the SU(3) symmetry parameter space which can be traversed using either the α_V and z SU(3) parameters or the vector meson–hyperon couplings as they are related by (4.60)–(4.62). For α_V and z to be uniquely determined the mixing angle must first be fixed, so for the analysis to follow in this section we assume ideal mixing ($\theta_{V,\text{ideal}} = 35.26°$).

To begin we will retain the assumption that the $S = -1$ Λ and Σ hyperons couple equally to the vector meson ($\alpha_V = 1$) and allow z to vary to cover the spectrum of possible values of $x_{\omega\Lambda} = x_{\omega\Sigma}$ and $x_{\omega\Xi}$ consistent with the SU(3) relations (4.60)–(4.62). The NS maximum mass and maximum strangeness fraction as a function of the vector couplings and z are given in Figs. 4.17 and 4.18 respectively. The minimum couplings and maximum z required for the satisfaction of the strong mass constraint along with the couplings and z values

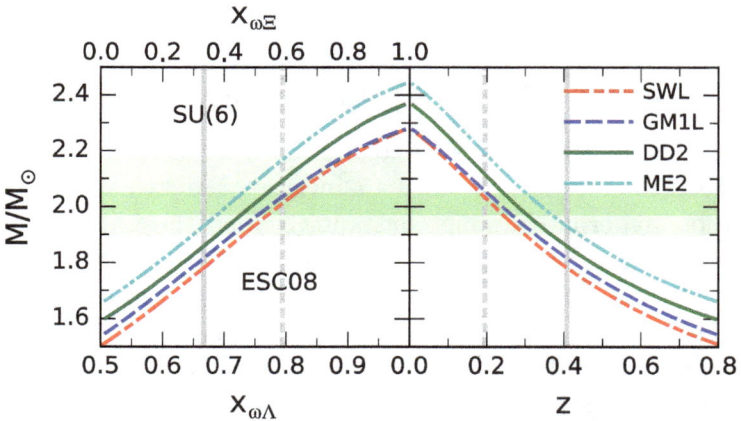

Fig. 4.17   The maximum NS mass as a function of the vector meson–hyperon coupling constants ($x_{\omega\Lambda}$ and $x_{\omega\Xi}$) and the SU(3) parameter z for $\alpha_V = 1$.

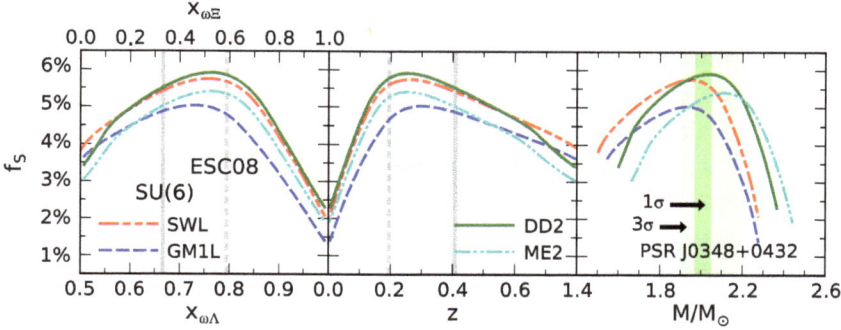

Fig. 4.18 The maximum strangeness fraction f_S as a function of the vector meson–hyperon coupling constants ($x_{\omega\Lambda}$ and $x_{\omega\Xi}$), the SU(3) parameter z, and the maximum mass, for $\alpha_V = 1$.

Table 4.17 The table shows the maximum coupling ratio z and minimum vector meson–hyperon coupling constants necessary for the satisfaction of the 2.01 $M_\odot$ mass constraint, as well as the maximum strangeness fraction f_S (in percent) and the associated coupling ratio z and vector meson–hyperon coupling constants. Note: $\alpha_V = 1$, $\theta_V = 35.26°$, and $x_{\omega\Xi} = x_{\omega\Lambda}$.

| $\alpha_V = 1$ | 2.01 $M_\odot$ | | | f_S^{max} | | | |
Parameterization	z	$x_{\omega\Lambda}$	$x_{\omega\Xi}$	z	$x_{\omega\Lambda}$	$x_{\omega\Xi}$	f_S^{max}
SWL	0.213	0.793	0.586	0.263	0.756	0.512	5.73%
GM1L	0.234	0.778	0.555	0.300	0.732	0.462	5.02%
DD2	0.274	0.748	0.497	0.254	0.762	0.525	5.91%
ME2	0.336	0.708	0.417	0.254	0.762	0.525	5.40%

associated with the maximum strangeness fraction are provided in Table 4.17. The maximum mass increases monotonically as the couplings tend toward universal and $z \to 0$. RMFL parameterizations require nearly equivalent couplings to that of the ESC08 model to satisfy the strong mass constraint, while the DDRMF parameterizations satisfy it for significantly lower couplings. The strangeness fraction increases with mass up to a maximum around $x_{\omega\Lambda} = 0.75$, $x_{\omega\Xi} = 0.5$, and $z = 0.25$, the maximum occuring at lower couplings and higher z than in the ESC08 model. Note that these couplings are almost perfectly situated at the halfway point between the minimum and maximum couplings allowed in SU(3) symmetry. DDRMF parameterizations with maximum strangeness satisfy the strong mass constraint, while the RMFL parameterizations satisfy only the weak mass constraint (right panel of Fig. 4.18). An analysis performed with standard RMF models found a very similar relationship between the maximum mass and strangeness fraction as that found in the right panel of Fig. 4.18 [Weissenborn, Chatterjee, and Schaffner-Bielich (2012a, 2014)].

The critical density for the onset of hyperonization is given in Fig. 4.19 as a function of the vector meson–hyperon coupling constants and the z parameter for $\alpha_V = 1$. Hyperons appear between $2.1 - 3.3\,n_0$ for all parameterizations, and the DDRMF critical densities are consistently between $0.2 - 0.3\,n_0$ less than those of the RMFL parameterizations. The lowest mass, charge and isospin neutral Λ is the first hyperon to appear for all couplings. The minimum critical densities that satisfy the strong mass constraint are $n_c^{\mathrm{SWL}} = 2.71\,n_0$, $n_c^{\mathrm{GM1L}} = 2.67\,n_0$, $n_c^{\mathrm{DD2}} = 2.42\,n_0$, and $n_c^{\mathrm{ME2}} = 2.32\,n_0$. In the following analysis we will allow α_V to vary as well, but it has been verified that the hyperon critical densities vary unremarkably with changes in α_V, so this analysis will not be duplicated.

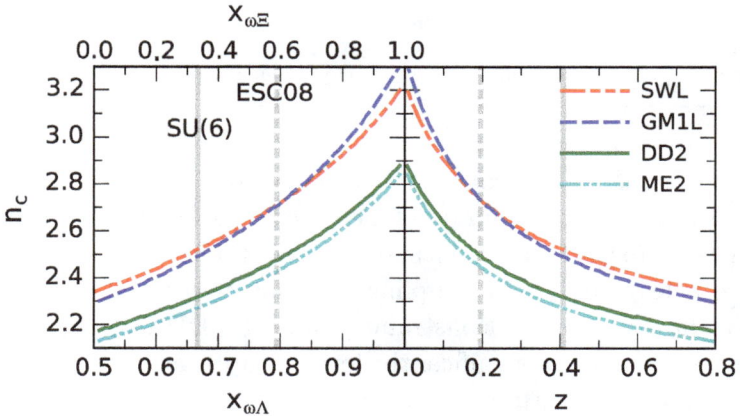

Fig. 4.19   The critical density for hyperonization n_c as a function of the vector meson–hyperon coupling constants ($x_{\omega\Lambda}$ and $x_{\omega\Xi}$) and the SU(3) parameter z for $\alpha_V = 1$.

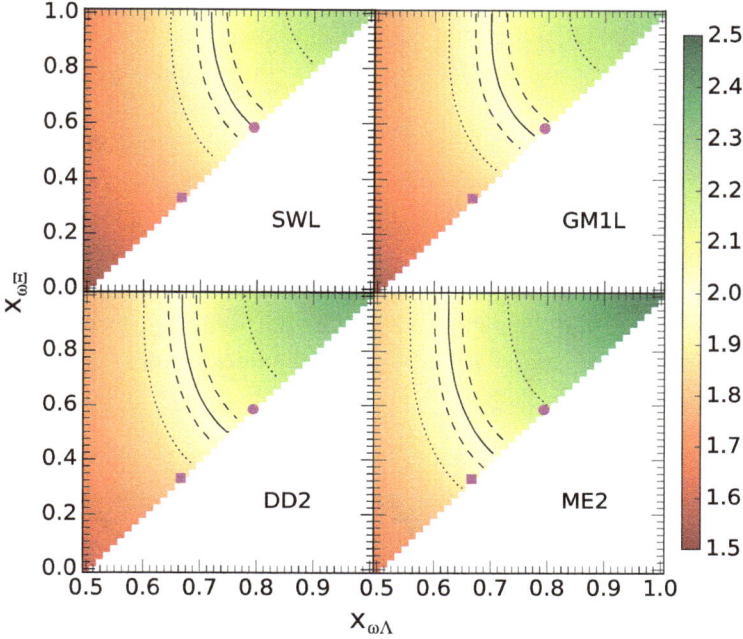

Fig. 4.20   Maximum mass in units of $M_\odot$ in the vector meson–hyperon coupling space. The solid black line indicates the 2.01 $M_\odot$ mass of PSR J0348+0432, with the black dashed (dotted) lines indicating the 1σ (3σ) uncertainty limits on that mass. The magenta square and circle markers indicate the vector meson–hyperon couplings determined in SU(6) symmetry and in the SU(3) ESC08 model with $\alpha_V = 1$. No interpolation has been employed.

To explore the entire vector meson–hyperon coupling space or SU(3) parameter space we must now relax the assumption that the Λ and Σ ($S = -1$) hyperons couple equally to the vector meson and allow α_V to vary between 0 and 1. The coupling ratio z varies in the range $0 \leq z \leq 2/\sqrt{6}$, bounded from above by the requirement that none of the vector meson–hyperon couplings should be negative. Figure 4.20 shows the maximum mass of NSs in the vector meson–hyperon coupling space. The solid contour in the figure represents the 2.01 $M_\odot$ mass of PSR J0348+0432 which constrains the couplings for the chosen parameterizations [Antoniadis *et al.* (2013)]. Larger values of the coupling constants consistently result in a greater maximum mass up to the maximum for the universal couplings $x_{\omega\Lambda} = x_{\omega\Sigma} = x_{\omega\Xi} = 1$. The SU(3) symmetry does not allow high values of $x_{\omega\Lambda}$ for low values of $x_{\omega\Xi}$, and as a result the mass is always much more sensitive to changes in $x_{\omega\Lambda}$ (moving horizontally in the space) than in $x_{\omega\Xi}$ (moving vertically). Treating the stiff ME2 parameterization as a limiting case, satisfaction of the strong mass constraint requires a minimum $x_{\omega\Lambda} \gtrsim 0.625$ and $x_{\omega\Xi} \gtrsim 0.4$.

The strangeness fraction for the vector meson–hyperon coupling space is given in Fig. 4.21. The maximum strangeness fractions, associated maximum masses, and coordinates in the vector meson–hyperon coupling space and SU(3) parameter space are given in Table 4.18. Hyperons appear for all couplings consistent with the

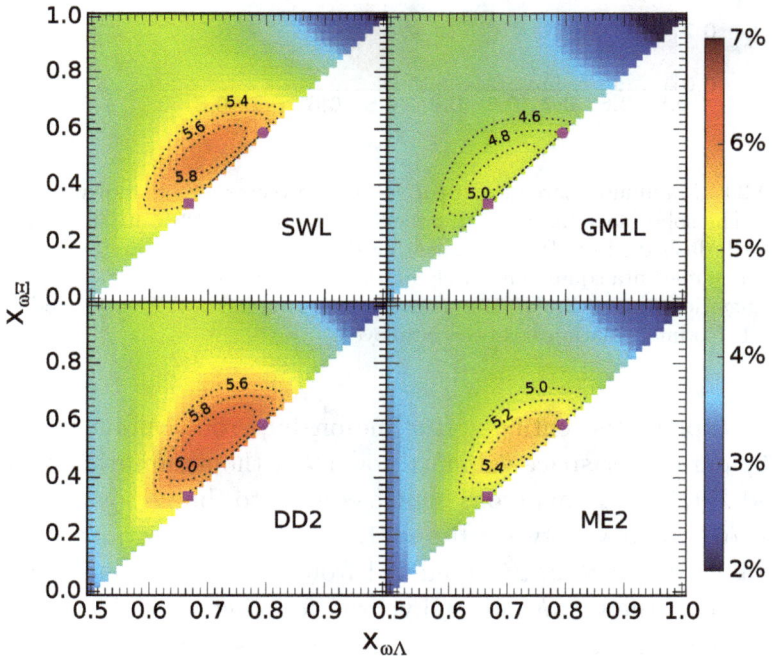

Fig. 4.21   Strangeness fraction f_S of the maximum mass NS in the vector meson–hyperon coupling space. The magenta square and circle markers indicate the vector meson–hyperon couplings determined in SU(6) symmetry and in the SU(3) ESC08 model with $\alpha_V = 1$. No interpolation has been employed.

Table 4.18   Maximum strangeness fraction of parameterizations in Figs. 4.21 and 4.22 along with the associated SU(3) coupling parameters, vector meson–hyperon coupling constants, and maximum mass value.

Parameterization	α_V	z	$x_{\omega\Lambda}$	$x_{\omega\Sigma}$	$x_{\omega\Xi}$	$M_{\text{max}}/M_\odot$	f_S^{max}
SW	0.350	0.375	0.755	1.130	0.697	1.99	5.69%
GM1	0.450	0.350	0.756	1.039	0.654	2.02	4.57%
SWL	0.550	0.425	0.699	0.957	0.526	1.90	5.96%
GM1L	0.675	0.425	0.685	0.859	0.457	1.89	5.12%
DD2	0.525	0.400	0.716	0.979	0.564	2.01	6.20%
ME2	0.600	0.375	0.723	0.924	0.546	2.08	5.57%

SU(3) symmetry, with a strangeness fraction $f_S \gtrsim 4\%$ for most of the coupling space for all parameterizations. Regions of high strangeness are clustered in hotspots in roughly the same location in the coupling space for the different RMFL and DDRMF parameterizations (these regions are outlined by dashed contours for emphasis in Fig. 4.21). These hotspots are slightly displaced from the $\alpha_V = 1$ diagonal where $x_{\omega\Lambda}$ is a maximum for a given value of $x_{\omega\Xi}$ favoring the presence of the doubly strange Ξ^-. Overall a higher strangeness fraction does not necessarily produce a lower maximum mass, which is made clear by tracing the $2.01\,M_\odot$ contour over the strangeness fraction. However, the lowest strangeness fractions are consistently found where the values of $x_{\omega\Lambda}$ and $x_{\omega\Xi}$ approach the universal coupling consistent with the region of highest maximum mass. The softer RMFL (SWL) and DDRMF (DD2) parameterizations present with higher strangeness fractions than their stiffer counterparts for the entirety of the coupling space, which is due in part to the higher core density reached by these parameterizations (see Table 4.14). The DD2 parameterization produces the largest strangeness fraction at 6.2% with an associated mass of $2.01\,M_\odot$, just satisfying the strong mass constraint.

The presence of strangeness hotspots is not limited to EoS with density-dependent meson–baryon couplings. Standard RMF parameterizations also exhibit strangeness hotspots in the vector meson–hyperon coupling space as shown in Fig. 4.22. The SW

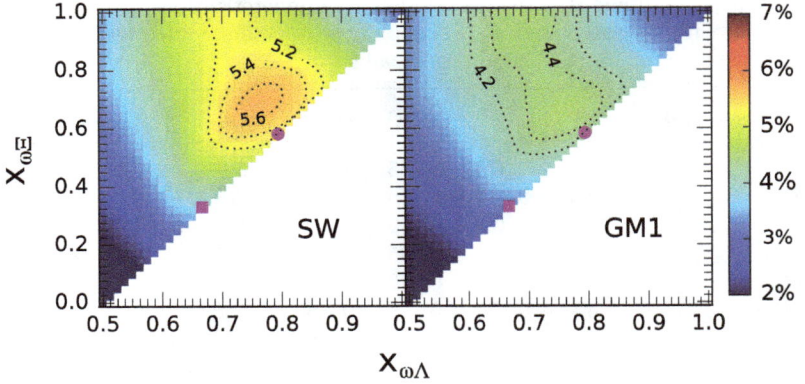

Fig. 4.22   Strangeness fraction of the maximum mass NS in the vector meson–hyperon coupling space for the standard RMF parameterizations SW and GM1. The magenta square and circle markers indicate the vector meson–hyperon couplings determined in SU(6) symmetry and in SU(3) symmetry with $\alpha_V = 1$. No interpolation has been employed.

parameterization has a well defined strangeness hotspot centered around (0.7625,0.700), but these coordinates and $x_{\omega\Xi}$ in particular have much higher values than those of SWL. The density dependence of the isovector meson–baryon coupling constant in the SWL and GM1L parameterizations shifts the strangeness landscape of standard RMF to lower values of $x_{\omega\Lambda}$ and $x_{\omega\Xi}$. In addition, the absence of density-dependent scalar and vector meson–baryon coupling constants in SWL and GM1L, and the similarity in the strangeness distributions to the DDRMF parameterizations, suggest that the density dependence of the isovector meson–baryon coupling constant and the slope of the asymmetry energy have a substantial effect on strangeness in NSs.

4.4.4.   *Varying the hypernuclear potentials*

The scalar meson–hyperon coupling constants are typically determined by fitting to hypernuclear potentials derived from experimental data as discussed in Sec. 4.4.1.2. The potential of the Λ hyperon is well established, however, a great deal of uncertainty still exists in the Σ and Ξ hyperon potentials due to either a lack of experimental data or ambiguous results. By varying the Σ and Ξ hyperon potentials their affect on the maximum mass and strangeness fraction of NSs can be analyzed. The vector meson–hyperon coupling constants are fixed at their SU(3) ESC08 values for this analysis (see Table 4.12).

The maximum mass of NSs in the Σ–Ξ hypernuclear potential space is given in Fig. 4.23, with $U_\Lambda^{(N)}$ fixed at -28 MeV. An increase in potential is accompanied by a decrease in the scalar meson–hyperon coupling constant for the given hyperon. The distribution of mass contours is extremely similar for all four parameterizations, but the maximum mass appears to be slightly more sensitive to potential changes the softer the EoS, the SWL contours being the most densely packed. The maximum mass is practically unchanged for $U_\Sigma^{(N)} \gtrsim 15$ MeV. However, at $U_\Sigma^{(N)} \lesssim -15$ MeV the maximum mass is very sensitive to changes in the Σ potential. Figure 4.24 shows the strangeness fraction of maximum mass NSs in the Σ–Ξ hypernuclear potential space. There is a clear difference in the way a decreasing Σ potential affects the strangeness in partially and fully density-dependent parameterizations. As long as $U_\Sigma^{(N)} \gtrsim 15$ MeV any change has little to no effect on the strangeness fraction for SWL and GM1L, and the same is true for DD2 and ME2 if $U_\Sigma^{(N)} \gtrsim 40$ MeV. However, the strangeness fraction is much more sensitive to changes

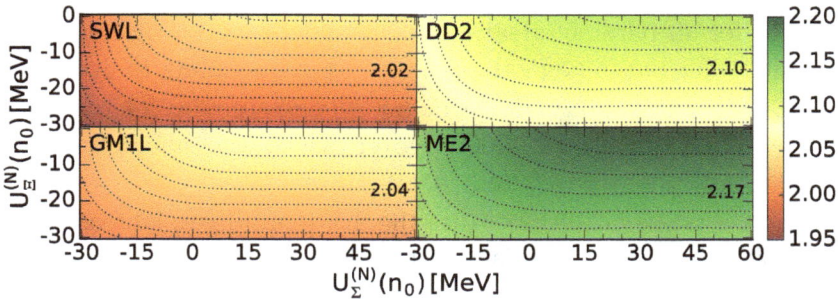

Fig. 4.23   Maximum mass in units of $M_\odot$ in the Σ–Ξ hypernuclear potential space. The black dotted lines are $0.01\,M_\odot$ spaced contours. No interpolation has been employed.

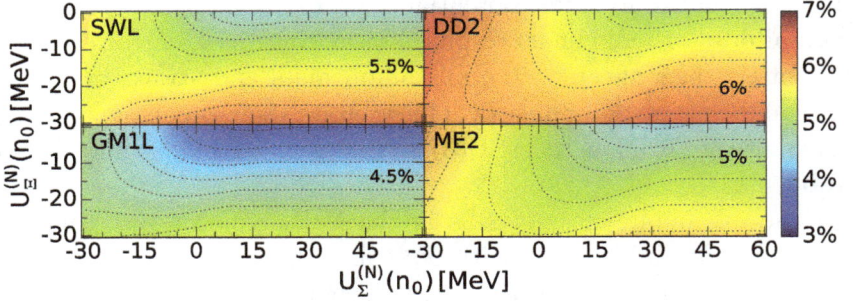

Fig. 4.24   Strangeness fraction of the maximum mass NS in the Σ–Ξ hypernuclear potential space. The black dotted lines represent 0.5% spaced contours. No interpolation has been employed.

in an attractive Σ potential in the DD2 and ME2 parameterizations largely due to the smaller isovector meson field shown in Fig. 4.15, indicating that an attractive Σ^- would be strongly favored over the heavier Ξ^-.

The Ξ potential dependence of the maximum mass and strangeness fraction can be analyzed by examining a cross section of Figs. 4.23 and 4.24 at the currently accepted value of the Σ potential $U_\Sigma^{(N)} = +30\,\mathrm{MeV}$ as shown in Fig. 4.25. Both the maximum mass and strangeness vary approximately linearly across the given range of $U_\Xi^{(N)}$. For a 1 MeV increase in potential the maximum mass increases by $\sim 0.0012\,M_\odot$ and the strangeness fraction decreases by $\sim 0.05\%$ for all parameterizations (slopes become slightly more shallow for stiffer parameterizations). The yellow shading indicates the $16 \pm 2\,\mathrm{MeV}$ uncertainty range in $U_\Xi^{(N)}$ suggested by [Oertel *et al.* (2015); Khaustov *et al.* (2000)], over which the maximum mass changes only slightly by about $0.005\,M_\odot$. Setting $U_\Xi^{(N)} = -18\,\mathrm{MeV}$ is the conservative choice because the maximum mass increases with increasing $U_\Xi^{(N)}$. Finally, it is clear that a rather substantial change in $U_\Xi^{(N)}$ would still not hinder satisfaction of the strong mass constraint by the DD2 or ME2 parameterizations.

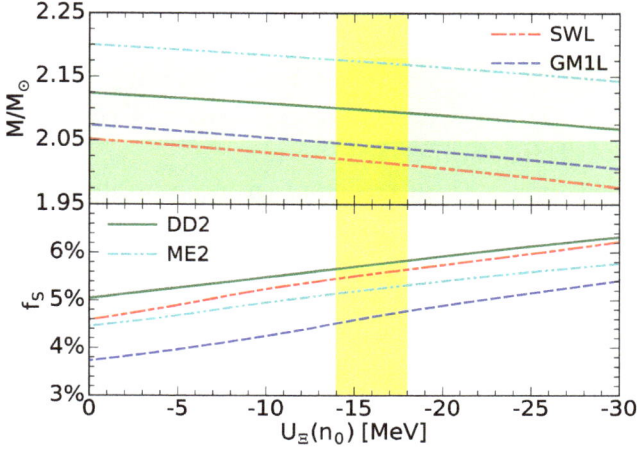

Fig. 4.25 Maximum mass and strangeness fraction as a function of the Ξ hypernuclear potential. The Λ and Σ potentials are fixed at $U_\Lambda^{(N)} = -28\,\text{MeV}$ and $U_\Sigma^{(N)} = +30\,\text{MeV}$.

4.5. Summary and Conclusions

In this work we studied constrained NS EoS considering hadronic matter that is both purely nucleonic as well as hyperonic in nature. To begin we introduced the relativistic mean-field (RMF) approximation to model the EoS of hadronic NS matter. This included improvements to the standard RMF approach such as the inclusion of density-dependent meson–baryon coupling constants (RMFL and DDRMF). We introduced the SW(L) parameterization created for this work as well as a number of other popular RMF and DDRMF parameterizations chosen from the literature. We found that standard RMF parameterizations such as SW, GM1, GM3, and NL3, are incompatible with numerous constraints as follows:

- The slope of the asymmetry energy (L_0) at the nuclear saturation density (n_0) in SNM is much higher than the upper limit of current expectations ($L_0 \lesssim 70\,\text{MeV}$).

- The nucleonic direct Urca (DU) process operates in NSs with masses that are much lower than the $M_{DU} \geq 1.5\,M_\odot$ constraint.
- The baryonic mass at a gravitational mass of $M_G = 1.249 \pm 0.001\,M_\odot$ is drastically inconsistent with the constraints determined from PSR J0737-3039.
- The canonical radius is inconsistent with composite determinations from NS observations and nuclear experiment/theory.

These issues were largely resolved for the SW, GM1, and GM3 parameterizations by introducing a density-dependent isovector meson–baryon coupling constant $(g_{\rho B}(n))$ with an additional parameter fit to an asymmetry energy slope of $L_0 = 55\,\mathrm{MeV}$ at n_0 in SNM (making them SWL, GM1L, and GM3L). Finally, it was determined that the SWL, GM1L, GM3L, IUFSU, TW99, DDF, DD2, and ME2 parameterizations were acceptable for use in the investigation of exotic matter in the NS core to follow.

Our investigation of exotic matter in NSs focused on hyperons and the determination of the meson–hyperon coupling constants. Initially the vector meson–hyperon coupling constants were set using either the SU(6) symmetry or the SU(3) ESC08 model, while the scalar meson–hyperon coupling constants were fit to empirical hypernuclear potentials. We found that employing SU(6) vector couplings led to hyperonic EoS that were too soft to satisfy the $2.01\,M_\odot$ strong mass constraint from PSR J0348+0432. However, employing the SU(3) ESC08 model vector couplings produced sufficiently stiff hyperonic EoS with the SWL, GM1L, DD2, and ME2 parameterizations. The GM3L, IUFSU, TW99, and DDF parameterizations with ESC08 couplings failed to satisfy even the weak mass constraint by a considerable margin and were discarded from further consideration.

Proceeding with the vector meson–hyperon coupling constants given by the SU(3) ESC08 model we found that hyperons first appear between $2.5 - 3\,n_0$. The Λ is the first hyperon to appear in all parameterizations followed by the relatively heavy Ξ^- in SWL and GM1L and the Σ^- in DD2 and ME2. However, the repulsive potential of the Σ^- causes it to quickly vanish replaced by the heavier Ξ^-. The hyperonic DU processes were shown to operate shortly after the appearance of the associated hyperons but at masses consistent with the $M_{DU} \geq 1.5\, M_\odot$ constraint.

Next we investigated the parameter space of the vector meson–hyperon coupling constants consistent with the SU(3) symmetry. We calculated the maximum mass and strangeness fraction of NSs on a finely-grained grid in the vector coupling space. The strong mass constraint was displayed as a contour on the heat-map of the maximum mass in the SU(3) vector meson–hyperon coupling space constraining the vector couplings. The strangeness fraction was found to be greater than 1% for the entire coupling space, and greater than about 4% for the majority of the space, indicating that hyperons appear for all vector couplings consistent with the SU(3) symmetry. We found that the maximum strangeness fractions are found in hotspots that form away from the $\alpha_V = 1$ diagonal, and that these hotspots are located in roughly the same region of the coupling space for the four remaining parameterizations, suggesting a strong link between the strangeness fraction and the asymmetry energy. The maximum strangeness fraction was found to be 6.2% at a mass of 2.01 $M_\odot$ in the DD2 parameterization.

Finally we fixed the vector meson–hyperon coupling constants with the SU(3) ESC08 model and varied the Ξ and Σ potentials to study the affect on the maximum mass and strangeness fraction. We found that as long as $U_\Sigma^{(N)} \gtrsim 15\,\text{MeV}$ it has very little effect on the mass and strangeness of NSs, and a 10 MeV change in $U_\Xi^{(N)}$ results in about a 0.02 $M_\odot$ change in maximum mass and about a half a percent change in strangeness.

Acknowledgments

This work is supported through the National Science Foundation under grant PHY-1714068.

References

Alford, M.G. *et al.*, *Rev. Mod. Phys.* **80**, 1455 (2008).
Ambartsumyan, V.A. and Saakyan, G.S., *Soviet Ast.* **4**, 187 (1960).
Antoniadis, J. *et al.*, *Science* **340**, 6131 (2013).

Baade, W. and Zwicky, F., *Phys. Rev.* **45**, 138 (1934).

Baym, G., Pethick, C.J., and Sutherland, T.P., *Astrophys. J.* **170**, 299 (1971).

Beringer, J. *et al.*, *Phys. Rev. C* **86**, 010001 (2012).

Blaschke, D., Grigorian, H., and Voskresensky, D.N., *Astron. Astrophys.* **424**, 979 (2004).

Bogdanov, S., *Astrophys. J.* **762**, 96 (2013).

Boguta, J. and Bodmer, A.R., *Nucl. Phys. A* **292**, 413 (1977).

Cavagnoli, R., Menezes, D.P., and Providência, C., *Phys. Rev. C* **84**, 065810, (2011).

Chadwick, J., *Nature* **129**, 312 (1932).

Chatterjee, D. and Vidana, I., *Eur. Phys. J. A* **52**, 29 (2016).

Danielewicz, P., Lacey, R., and Lynch, W.G., *Science* **298**, 1592 (2002).

Danielewicz, P. and Lee, J., *Nucl. Phys. A* **922**, 1 (2014).

Demorest, P. *et al.*, *Nature* **467**, 1018 (2010).

Douchin, F. and Haensel, P., *Astron. Astrophys.* **380**, 151 (2001).

Dover, C.B. and Gal, A., *Prog. Part. Nucl. Phys.* **12**, 171 (1984).

Drago, A. *et al.*, *Phys. Rev. C* **90**, 065809 (2014).

Drago, A., Lavagno, A., and Pagliara, G., *Phys. Rev. D* **89**, 043014 (2014).

Drago, A. *et al.*, *Eur. Phys. J. A* **52**, 40 (2016).

Drago, A. and Pagliara, G., *Eur. Phys. J. A* **52**, 41 (2016).

Dutra, M. *et al.*, *Phys. Rev. C* **85**, 035201 (2012).

Dutra, M. *et al.*, *Phys. Rev. C* **90**, 055203 (2014).

Dutra, M., Lourenço, O., and Menezes, D., *Phys. Rev. C* **93**, 025806 (2016).

Fonseca, E. *et al.*, *Astrophys. J.* **832**, 167 (2018).

Fortin, M. *et al.*, *Phys. Rev. C* **95**, 065803 (2017).

Fuchs, C., Lenske, H., and Wolter, H.H., *Phys. Rev. C* **52**, 3043 (1995).

Fuchs, C., *Prog. Part. Nucl. Phys.* **56**, 1 (2006).

Furnstahl, R.J., Rusnak, J.J., and Serot, B.D., *Nucl. Phys. A* **632**, 607 (1998).

Glendenning, N.K., in *Compact Stars: Nuclear Physics, Particle Physics, and General Relativity* (Springer-Verlag, New York, 1997).

Glendenning, N.K., *Astrophys. J.* **2930**, 470 (1985).

Gomes, R.O. *et al.*, *Astrophys. J.* **808**, 8 (2015).

Haensel, P. and Pichon, B., *Astron. Astrophys.* **283**, 313 (1994).

Haensel, P., Potekhin, A.Y., and Yakovlev, D.G., in *Neutron Stars 1: Equation of State and Structure* (Springer Verlag, New York, 2007).

Hagen, G. *et al.*, *Nat. Phys.* **12**, 186 (2015).

Hambaryan, V. *et al.*, *J. Phys. Conf. Ser.* **496**, 012105 (2014).

Hessels, J.W.T. *et al.*, *Science* **311**, 1901 (2006).

Horowitz, C.J. and Piekarewicz, J., *Phys. Rev. C* **64**, 062802 (2001).

Horowitz, C.J. and Piekarewicz, J., *Phys. Rev. Lett.* **86**, 5647 (2001).

Ivanenko, D.D. and Kurdgelaidze, D.F., *Astrophysics* **1**, 251 (1965).

Jaminon, M. and Mahaux, C., *Phys. Rev. C* **40**, 354 (1989).

Johnson, C.H., Horen, D.J., and Mahaux, C., *Phys. Rev. C* **36**, 2252 (1987).

Khaustov, P. *et al.*, *Phys. Rev. C* **61**, 054603 (2000).

Kitaura, F.S., Janka, H.-Th., and Hillebrandt, W., *Astron. Astrophys.* **450**, 345 (2006).

Klahn, T. *et al.*, *Phys. Rev. C* **74**, 035802 (2066).

Kolomeitsev, E.E. and Maslov, K.A. and Voskresensky, D.N., *Nucl. Phys. A* **961**, 106 (2017).

Kramer, M. *et al.*, *eConf* C**041213**, 0038 (2004).

Lalazissis, G.A. *et al.*, *Phys. Rev. C* **71**, 024312 (2005).

Lattimer, J.M., *Universe* **5**(7), 159 (2019).

Lattimer, J.M. *et al.*, *Phys. Rev. Lett.* **66**, 2701 (1991).

Lattimer, J.M. and Lim, Y., *Astrophys. J.* **771**, 51 (2013).

Lattimer, J.M. and Steiner, A.W., *Eur. Phys. J. A* **50**, 40 (2014).

Li, B.-A. *et al.*, *Phys. Rep.* **464**, 113 (2008).

Lim, Y., Lee, C.-H., and Oh, Y., *Phys. Rev. C* **97**, 023010 (2018).

Lopes, L.L. and Menezes, D.P., *Phys. Rev. C* **89**, 025805 (2014).

Lynch, W.G. *et al.*, *Prog. Part. Nucl. Phys.* **62**, 427 (2009).

Mahaux, C. and Sartor, R., *Nucl. Phys. A* **475**, 247 (1987).

Maslov, K.A., Kolomeitsev, E.E., and Voskresensky, D.N., *Nucl. Phys. A* **950**, 64 (2016).

Miyatsu, T., Cheoun, M., and Saito, K., *Phys. Rev. C* **88**, 015802 (2013).

Moller, P. and Nix, J.R., *Nucl. Phys. A* **361**, 117 (1981).

Müller, H. and Serot, B.D., *Nucl. Phys. A* **606**, 508 (1996).

Oertel, M., Providencia, C., Gulminelli, F., and Raduta, Ad. R., *J. Phys. G* **42**, 075202 (2015).

Olausen S.A. and Kaspi, V.M., *Astrophys. J. Suppl.* **212**, 6 (2014).

Oppenheimer, J.R. and Volkoff, G.M., *Phys. Rev.* **55**, 374 (1939).

Orsaria, M. *et al.*, *Phys. Rev. C* **89**, 015806 (2014).

Özel, F., Baym, G., and Güver, T., *Phys. Rev. D* **82**, 101301 (2010).

Patrignani, C. *et al.*, *Chin. Phys. C* **40**, 100001 (2016).

Podsiadlowski, P. *et al.*, *Mon. Not. Roy. Astron. Soc.* **361**, 1243 (2005).

Popov, S. *et al.*, *Astron. Astrophys.* **448**, 327 (2006).

Prakash, M. *et al.*, *Astrophys. J.* **390**, L77 (1992).

Providência, C. and Rabhi, A., *Phys. Rev. C* **87**, 055801, (2013).

Providência, C. *et al.*, *Eur. Phys. J. A* **50**, 44 (2014).

Reardon, H.J. *et al.*, *Mon. Not. Roy. Astron. Soc.* **455**, 1751 (2016).

Rijken, T.A., Nagels, M.M., and Yamamoto, Y., *Prog. Theor. Phys. Suppl.* **185**, 14 (2010).

Schaffner, J. *et al.*, *Ann. Phys.* **235**, 35 (1994).

Schaffner, J. and Gal, A., *Phys. Rev. C* **62**, 034311 (2000).

Shlomo, S., Kolomietz, V.M., and Colo, G., *Eur. Phys. J. A* **30**, (2006).

Steiner, A.W., Lattimer, J.M., and Brown, E.F., *Astrophys. J.* **722**, 33 (2010).

Stone, J.R., Stone, N.J., and Moszkowski, S.A., *Phys. Rev. C* **89**, 044316 (2014).

The Super Proton Synchrotron at CERN, https://home.cern/about/accelerators/super-proton-synchrotron/ (2018).

The Facility for Antiproton and Ion Research at GSI, http://www.fair-center.eu/ (2018).

The Large Hadron Collider at CERN, https://home.cern/topics/large-hadron-collider (2018).

The Nuclotron-based Ion Collider fAcility at JINR, http://nica.jinr.ru/ (2018).

The Relativistic Heavy Ion Collider at BNL, https://www.bnl.gov/rhic/ (2018).

Thew, I., Lattimer, J.M., Ohnishi, A. and Kolomeitsev, E.E., *Astrophys. J.* **848**, 105 (2017).

Tolman, R.C., *Phys. Rev.* **55**, 364 (1939).

Tolos, L., Centelles, M., and Ramos, A., *Astrophys. J.* **834**, 3 (2017).

Typel, S. and Wolter, H.H., *Nucl. Phys. A* **656**, 331 (1999).

Typel, S., *Phys. Rev. C* **71**, 064301 (2005).

Typel, S. *et al.*, *Phys. Rev. C* **81**, 015803 (2010).

Weber, F., in *Pulsars as Astrophysical Laboratories for Nuclear and Particle Physics* (IoP Publishing, London, 1999).

Weber, F., *Prog. Part. Nucl. Phys.* **54**, 193 (2005).

Weissenborn, S., Chatterjee D., and Schaffner-Bielich, J., *Phys. Rev. C* **85**, 065802 (2012); *Phys. Rev. C* **90**, 019904(E) (2014).

Weissenborn, S., Chatterjee D., and Schaffner-Bielich, J., *Nucl. Phys. A* **881**, 62 (2012).

Weissenborn, S., Chatterjee D., and Schaffner-Bielich, J., *Nucl. Phys. A* **914**, 421 (2013).

Chapter 5

Probing the Spacetime Around a Black Hole with X-Ray Variability

Tomaso M. Belloni

INAF — Osservatorio Astronomico di Brera
Via E. Bianchi 46, I-23807 Merate, Italy
tomaso.belloni@inaf.it

In the past decades, the phenomenology of fast time variations of high-energy flux from black-hole binaries has increased, thanks to the availability of more sophisticated space observatories, and a complex picture has emerged. Recently, models have been developed to interpret the observed signals in terms of fundamental frequencies connected to General Relativity (GR), and have opened this way a promising way to measure prediction of GR in the strong-field regime. In this contribution, a review about the current standpoint from both the observational and theoretical points of view is presented. As a result of this analysis, these systems emerging as promising laboratories for testing GR and the observations available today suggest that the next observational facilities can lead to a breakthrough in the field.

5.1. Introduction: The Promise of X-Ray Binaries

X-ray binaries are stellar binary systems that contain a normal star and a collapsed object, either a black hole or a neutron star, in which the strong gravitational pull of the latter strips matter from the companion and accretes it. The process of accretion is complex and not fully understood. Because of its angular momentum, the gas cannot fall directly onto/into the compact object, but forms a disk around it. Matter at each radius follows a Keplerian orbit, slowly spiraling towards the center, due to friction, emitting radiation at higher and higher energies as it nears the collapsed object. The inner parts of the accretion flow, below $\sim$100 R_g, are very hot and strong X-ray emission is observed. The structure of the accretion flow is very complex and varies with time depending on the rate of mass flowing through it. Although the supply from the companion star is expected to be constant, instabilities in the accretion flow result in a variable accretion rate through the inner parts of the disk. This variability can be extreme in transient systems, in which the compact object spends most of its time accreting at a very low rate, to experience surges of accretion for periods of weeks to months when the observed high-energy luminosity increases by several orders of magnitude [Belloni and Motta (2016)].

An important aspect of the study of X-ray binaries is the determination of the nature of the compact object. While direct evidence for the presence of a neutron star can be obtained from the observation of high-frequency pulsations or thermonuclear X-ray bursts (caused by unstable nuclear burning on the surface of the object, indication that a surface is present), the black-hole nature of the central object has not so far been ascertained with direct measurements [Belloni (2018)]. The strongest indication for a black hole is indirect, namely the measurement of the mass of the compact object from studying the optical modulation from the binary system [Casares and Jonker (2014)].

To determine the presence of a black hole directly, we need to be able to observe in the X-ray emission, which originates very close to the black hole, effects due to General Relativity (GR), which would also allow us to check the validity of the theory in the strong field regime. The best approach would be to measure the mass of the compact object directly, with additional ideal observables being GR precession frequencies of orbiting matter and the presence of an innermost stable circular orbit (ISCO). The absence of a solid surface would be very difficult to prove as it is a negative measurement. For the identification and measurement of GR effects in the strong field, X-ray binaries are our best laboratories. Active galactic nuclei, which contain supermassive black holes, are not a match as their gravitational potential is equivalent, but the field curvature much less [Psaltis (2008)]. Spectacular measurements are being provided by the observation of the double pulsar [Kramer *et al.* (2008)]. However, the two neutron stars in the system, that can be treated as test particles, are 7×10^5 km apart. The same applies to the original Hulst & Taylor binary pulsar [Weisberg and Huang (2016)]. In the case of a black-hole binary, our test particles orbit the object at a few gravitational radii, but they are not really test particles, but plasma in a complex accretion flow whose properties are not completely known. Nevertheless, these laboratories are available thanks to X-ray astronomical satellites that provide high-energy data of increasing quality (and quantity) and are now giving us the first answer. In this chapter, I will concentrate on the first answers offered by the analysis of time variability in the X-ray flux, which close to the black hole takes place on time scales well below a second.

5.2. Spectral Approaches

Before introducing time variability, it is important to mention other methods that are being used to estimate GR parameters based on the analysis of emitted X-ray spectra [Miller and Miller (2014)]. These methods can be very powerful and are yielding more and more refined estimates of black-hole spins.

5.2.1. *Continuum spectra*

The energy spectra originating from black-hole binaries (hereafter BHB) are very complex, being the superposition of different components from different regions of the accretion flow, and time dependent. One component that is observed is produced in a geometrically thin and optically thick accretion disk, as modeled originally by Shakura and Sunyaev [Shakura and Sunyaev (1973)]. The spectrum emitted by such a disk is the superposition of black-body components from different radii, with a specific radial temperature distribution. The integrated spectrum has the luminosity of a sphere whose radius is the inner radius of the disk and whose temperature is that at the inner radius. Essentially, it behaves like a black body, which means that an estimate of its luminosity and temperature directly translates into a measurement of the inner radius of the disk. Given a black-hole mass (estimated from optical data), the radius can be expressed in gravitational radii and if it corresponds to the ISCO it yields the spin of the black hole. The emitted spectrum is more complex than a simple sum of black-body components and more sophisticated and realistic models have been produced [Davis and Hubeny (2006)].

Despite the (variable) presence of several other spectral components, BHBs in their so-called "high-soft state" emit a spectrum that is consistent with an almost pure thermal component from the accretion disk. Detailed modeling showed that the estimated inner disk radius is constant and can therefore be associated to the ISCO. This has led to the measurement of the black-hole spin for a number of sources [McClintock, Narayan, and Steiner (2014)]. In order to measure the spin one must assume that the measured inner radius corresponds to ISCO, but any larger radius would lead to a higher spin. However, in order to measure the radius one needs to model the spectrum very accurately, take care of possible weak contamination from additional components, assume a distance to the source and assume an inclination of the system (the disk is not a sphere and its luminosity depends on its inclination, which of course is a random parameter and can be estimated with optical measurements). The presence of interstellar and local absorption also complicates the measurement.

5.2.2. *Iron line emission*

A different and very powerful method used to estimate black-hole spins is related to an additional component of the energy spectrum. The inner region of the accretion disk emits radiation that also irradiates on the disk. The incident X-ray radiation is "reflected" by the disk and emits a spectrum that is composed of fluorescence lines, of which the most prominent are Fe K lines, photoelectric absorption from the Fe shell above the lines, and an increase in form of a continuum above that, with a gradual flux decrease at high energies. This spectrum is complex, but it is dominated by an iron emission line that in a cloud of gas would be very narrow. However, an accretion disk is made of matter orbiting at the local Keplerian period, which means that the overall spectrum will be the superposition of spectra from different annuli. The spectrum from each annulus will be modified by Doppler effect due to the fast rotation, by relativistic redshift and boosting. The final shape is expected to be very broad and possibly double-horned [Ross and Fabian (2007); Fabian and Ross (2010)]. Of course all these effects affecting the line depend on the inclination, but it turns out that the blue wing is sensitive mostly to the inclination and the red wing to how close the disk comes to the black hole because of the strong redshift. Therefore, detecting a broad line excess and modeling it with the proper model can potentially yield a measurement of ISCO. Several measurements have been obtained, in some cases combined with the continuum method described above [Steiner *et al.* (2011)]. This method must deal with the same issues as the one above, with the exception of the distance measurement, as radii are obtained directly in units of gravitational radii. In particular, the full energy spectrum extends over a broad X-ray band and is very complex, with several components some of which overlap in the energy region where the broad line is observed (6–7 keV). Since a very broad band extending from 3 keV to 7 keV is observationally a continuum component, this means that the full broad-band spectrum must be successfully fitted in order to estimate the line parameters.

5.3. Fast Time Variability

The X-ray emission observed from BHBs is very variable. Throughout the evolution of an outburst of a transient source, when accretion rate increases and X-ray luminosity reaches values of 10^{36-39} erg/s, different states are identified, which correspond to very different energy spectra and fast time variability. One of these states, the already mentioned high-soft state, when the emitted spectrum is almost a pure thermal component (that is used by the continuum methods), shows very little fast variability, but the others can see variability up to $\sim$140% fractional variability. This variability is in the form of broad-band noise and of peaked components called quasi-periodic oscillations (QPO). These peaks are broader than coherent oscillations, but yield very precise measurements of characteristic time scales. In Figs. 5.1 and 5.2 one can see two examples. In Fig. 5.1 there are four peaks in addition to noise, but they are harmonically related so only one frequency (that of the strongest peak) is usually considered. In Fig. 5.2, two peaks at higher frequency are seen, with low noise. They are also harmonically related.

It was clear from the first attempts of theoretical modeling of X-ray emission from X-ray binaries that variability could be produced by inhomogeneities in the inner accretion flow. If inhomogeneities can live longer than the orbital time scale at their radius, this could lead to a signal concentrated around the frequencies corresponding to that radius. In this way the inhomogeneity could be used as a "test particle" to probe the accretion flow and at the same

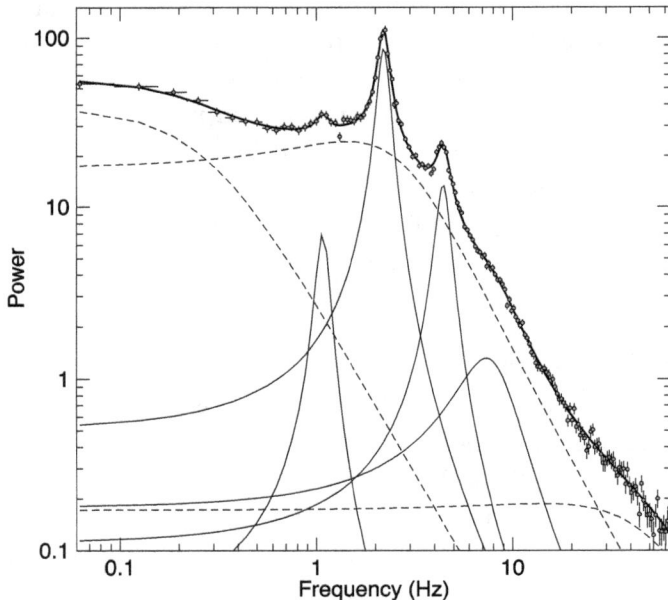

Fig. 5.1   Power density spectrum (PDS) from the source GRS 1915+105. Several low-frequency components can be seen. The frequencies of the four peaks are in harmonic relation [Ratti, Belloni, and Motta (2012)].

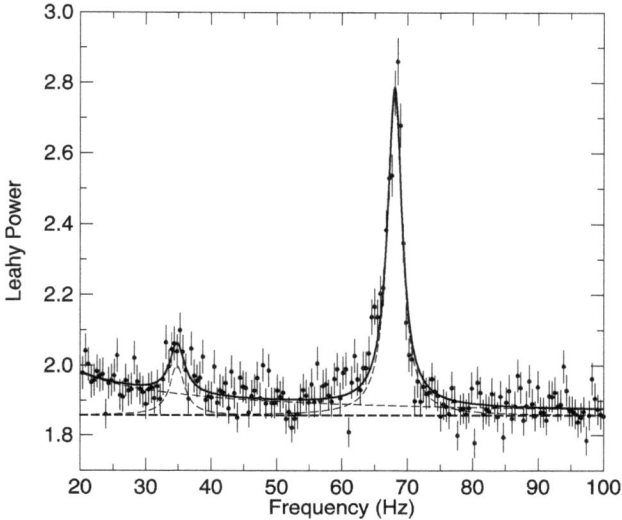

Fig. 5.2   Another PDS from GRS 1915+105 in a different state, where high-frequency QPOs are observed [Belloni and Altamirano (2013)]. Two peaks are observed here, with harmonically related frequencies.

time the spacetime close to a black hole. For instance observing high frequencies, the position (and existence) of the ISCO could be measured. After these ideas were put forward, QPOs were observed, first in neutron-star binaries then in BHBs. These provided precise frequencies to input to models, although the situation is complicated by the fact that an accretion flow is not made of tennis balls, but is a complex structure of orbiting plasma with its own characteristic time scales, which in principle could also lead to observable signals.

In the 1990s, multiple QPO peaks started being observed, thanks to NASA's RossiXTE satellite for X-ray astronomy, designed for analysis of fast variability [van der Klis (2006)]. Dealing with more than one time scale involves adding different frequencies in addition to the orbital ones. This led to the development of models based on fundamental frequencies of matter orbiting a collapsed object: in addition to the Keplerian one (ν_ϕ), the radial epicyclic frequency (ν_r) and the vertical epicyclic frequency (ν_θ) were considered. For a test particle orbiting at radius r on a slightly eccentric orbit slightly tilted from the plane perpendicular to the spin of a Kerr black hole with specific angular momentum J, these are:

$$\nu_\phi = \sqrt{GM/r^3}/2\pi \, (1 + a(r_g/r)^{3/2})$$
$$\nu_r^2 = \nu_\phi^2 \, [1 - 6(r_g/r) + 8a \, (r_g/r)^{3/2} - 3a^2 \, (r_g/r)^2]$$
$$\nu_\theta^2 = \nu_\phi^2 \, [1 - 4a \, (r_g/r)^{3/2} + 3a^2 \, (r_g/r)^2]$$

where $r_g = GM/c^2$ and $a = Jc/GM^2$. In addition, possible candidates are also the periastron precession frequency $\nu_{per} = \nu_\phi - \nu_r$ and the nodal precession frequency $\nu_{nod} = \nu_\phi - \nu_\theta$. In Newtonian approximation ν_ϕ, ν_r and ν_θ are identical, but in a strong gravitational field they are not: ν_ϕ and ν_θ decrease with radius, but ν_r is null at ISCO, has a maximum at a specific radius, then decreases at larger radii. Clearly, matching the observed frequencies with the values from these equations offers the possibility of testing important predictions of GR in the strong-field regime. Notice that the equations above depend only on three parameters: the mass and spin of the black hole and the radius of the orbit. A measurement of three frequencies, if they can be associated to these physical quantities, would give a direct measurement of mass and spin.

5.4. Observations

Going into the intricacy of source states would be beyond the scope of this chapter. I will present only the basic information that can be applied to theoretical models. Most of the variability that is observed takes place at low frequencies, which we can define as those below $\sim 10-20 \, \text{Hz}$, although the most interesting signals are those observed at higher frequencies, in the range where ν_ϕ and ν_r are expected for a stellar-mass black hole.

5.4.1. *Low frequencies*

When BHBs are in their hard state (when the thermal disk described above is not observed and the emission is dominated by other components), strong variability is observed in the form of noise, although

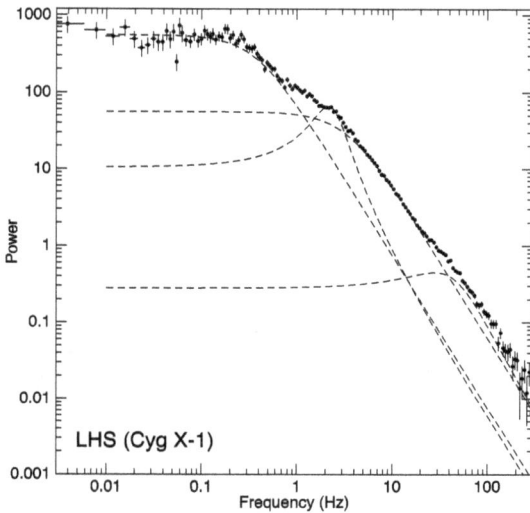

Fig. 5.3   PDS from the hard state of Cyg X-1. The dashed lines mark the different noise components.

sometimes a more peaked component, a QPO, can also be observed. The typical power density spectrum is that shown in Fig. 5.3. Four components are seen here, one of which slightly peaked. By applying a Lorentzian model for these components, it is possible to extract a characteristic frequency from each of them [Belloni, Psaltis, and van der Klis (2002)].

As the X-ray flux varies, these frequencies change in a correlated way. Recently, the lowest-frequency component has been associated with propagation effects in the accretion flow, where variability from all radii in the disk contributes to the noise [Arévalo and Uttley (2006); Ingram (2016)], although the frequency of this component is correlated with the others, an indication that the situation could be more complicated.

When the energy spectrum softens and a disk component starts appearing in the 2–10 keV band, the PDS is consistent with being a high-frequency extension of the one in Fig. 5.3, meaning that all frequencies are higher and the total fractional variability is lower. However, a clear peaked component appears, called type-C QPO

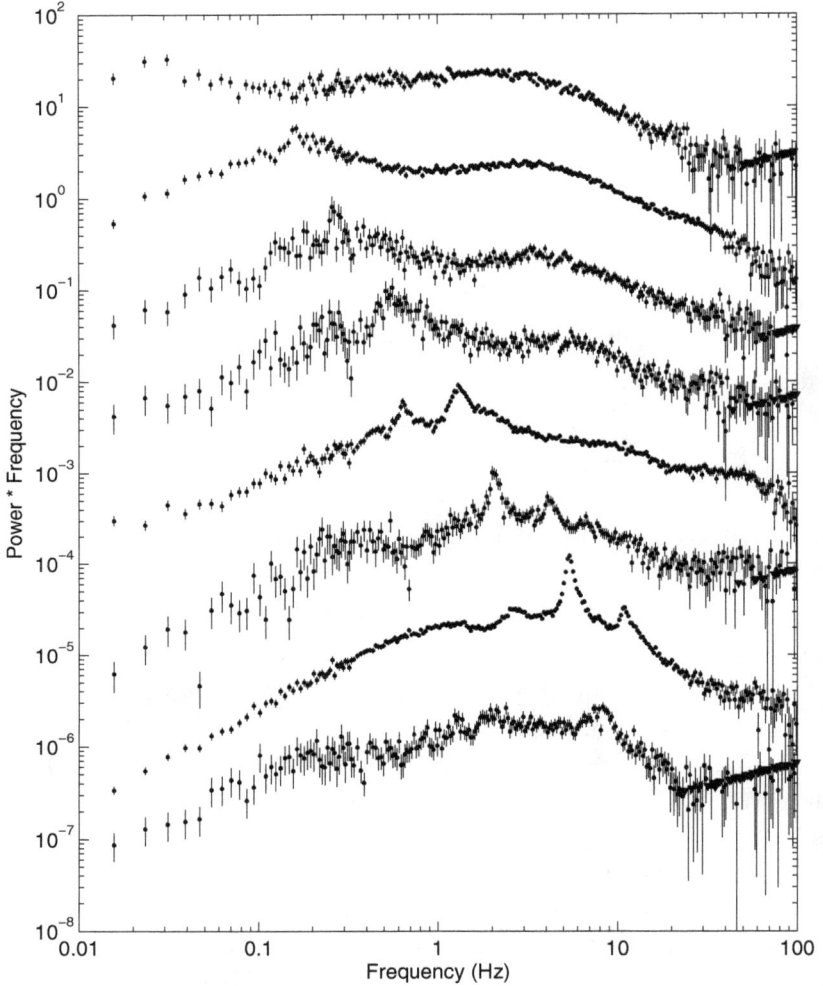

Fig. 5.4   Time sequence (from top to bottom) of PDS (multiplied by frequency) of GX 339-4. The evolution from noise to noise+QPO and the increasing frequencies is evident. From Belloni *et al.* (2008).

(see Figs. 5.1 and 5.4). As the name suggests, there are at least two other types of QPOs (type-A and B), but the important one to consider in this context is type-C, the most common one. These are QPO peaks whose centroid frequency varies in the 0.01–30 Hz range.

They are rather narrow, with a FWHM roughly a tenth of their centroid frequency and can be as strong as 5–10% fractional rms. They appear often with harmonically related peaks (see Fig. 5.1) and their centroid correlates with the energy spectrum, being highest when the spectrum is softest. Their energy spectrum is too hard for the modulated emission to originate from the thermal accretion disk, but the large changes in frequency suggest that the associated radius of emission changes. A strong correlation between their centroid frequency and the characteristic frequency of the lowest noise component has been discovered [Wijnands and van der Klis (1999)]. Type-C QPOs have been observed from many sources and appear to be very common. Their frequency is too low to be associated to Keplerian orbits, but it can be linked to precession. In softer states either other types of QPO are observed or, in the full-fledged high-soft state, there is little variability.

5.4.2. *High frequencies*

While signals at low frequencies are almost always observed, high-frequency features appear to be rare (or weak). Only with the RossiXTE mission we have started to sample efficiently this frequency range, which led to the discovery of the (related) kHz QPO in neutron-star binaries, of which there are many detections. For black holes, the situation is not so good. The first observation of a high-frequency QPO from a BHB was at 67 Hz from the very variable source GRS 1915+105 [Morgan, Remillard, and Greiner (1997)]. After that detection, in sixteen years of operation RXTE discovered very few more from other sources, although GRS 1915+105 showed many (see Table 5.1) [Belloni, Sanna, and Méndez (2012); Belloni and Altamirano (2013)]. Some reported detections are at low statistical significance and others were too broad to be classified as QPOs. We are left with *six* sources with at least a significant detection, see Table 5.1. All these correspond to observations at very high luminosity and therefore strong signal; it is unclear whether this is an effect of QPOs being stronger at high luminosity or our sensitivity being too low at lower luminosities.

Table 5.1   High-frequency QPOs from BHBs.

Source	N_{peaks}	Simultaneity	Frequency (Hz)
GRO J1655-40	2	Y	$\sim$300, $\sim$400
XTE J1550-564	2	N	$\sim$180, $\sim$280
XTE J1650-500	1	—	$\sim$250
H 1743-322	2	N	$\sim$160, $\sim$240
IGR 17091-3624	1	—	$\sim$66
GRS 1915+105	4	Y	$\sim$27, $\sim$34, $\sim$41, $\sim$67

Obviously these peaks are not as easy to detect as type-C QPOs, in addition they seem to be incompatible with the presence of type-C QPOs. They only appear in different states from that of type-C QPOs and there are only a few cases of simultaneous detection. As can be seen from Table 5.1, three sources (I will discuss the special case of GRS 1915+105 separately) have shown two peaks, although only in one case there are two significant detections in the same observation. For these three systems, the frequencies of the two peaks are close to being in a 3:2 ratio. For GRS 1915+105 things are, as usual, more complicated and no 3:2 ratio is present among the detections in Table 5.1, although other small integer ratios can be extracted.

The case of GRS 1915+105 is different. This system is very peculiar and, unlike normal transient systems, has been very bright throughout the whole RossiXTE period, which means that it has been observed many times at very high accretion. Indeed, a systematic search through the archive has lead to 51 detections of high-frequency QPOs, out of which 48 have a centroid frequency in the 63–71 Hz range [Belloni and Altamirano (2013)]. This frequency is a very stable number in this system. However, the most promising candidate for its interpretation, namely ν_ϕ at ISCO is not an option, as the current mass measurements predict a higher frequency even in the case of zero spin.

5.5. Models

The presence of noise in the X-ray emission of X-ray binaries was determined with the first observations from space with the *Uhuru* satellite, when relatively long observations could be obtained. The quality of the data did not allow precise determination of the noise properties and early models concentrated on a possible shot-noise nature of the variability [Terrell (1972); Nolan *et al.* (1981); Belloni and Hasinger (1990)], which we now know can be excluded due to statistical properties of the time series. The first QPO peaks in the 1980s were discovered at low frequencies from NS binaries and led to the development of models that involved the spin of the neutron star [Lamb *et al.* (1985)]. The discovery of QPOs from BHBs, where these models cannot be applied, changed little. However, the discovery of high-frequency features with RossiXTE in the second half of the 1990s changed everything and led to the development of models that could be applied to both classes of systems, independent of the nature of the compact object.

5.5.1. *The relativistic precession model*

The presence of three main features in the variability of neutron-star models and the availability of a large number of detections, led to the search for models that could explain all the QPOs. There are additional types of QPOs and sideband peaks, but it is the three peaks, one at low frequencies and two at high frequencies (called kHz QPOs because of their high frequency that can reach 1000 Hz) that can form the base for a successful model. The first such model is the relativistic precession model (RPM) [Stella and Vietri (1998, 1999); Stella, Vietri, and Morsink (1999)]. It is a local model, in the sense that the oscillations are associated to a specific radius in the accretion flow, and as most models it only aims at the interpretation of the observed frequencies and does not address a production mechanism. In the RPM, the low-frequency QPO is interpreted as ν_{nod} and the two high-frequency ones as ν_{per} and ν_ϕ. The frequencies are associated to a single radius around the neutron star. All three GR frequencies outlined above are involved. When applied to the available kHz QPO data, the model did not provide a statistically good fit, but yielded frequencies in the observed range and with the basic dependence between each other. Figure 5.5 shows the difference

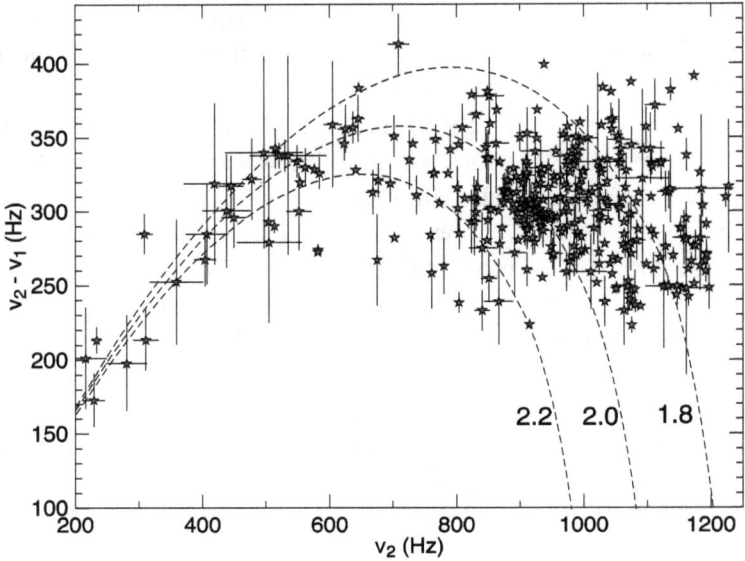

Fig. 5.5   Difference between the frequencies of the two kHz QPO peaks in NS binaries as a function of the upper frequency, where all published pairs of QPOs have been included. This plot is a new version of earlier published ones [Stella and Vietri (1999); Boutloukos *et al.* (2006)]. The lines correspond to RPM predictions for three values of the neutron-star mass.

between the frequencies of the two kHz QPO peaks (in the RPM corresponding to the radial epicyclic frequency ν_r) as a function of the upper frequency (corresponding to ν_ϕ) from all published pairs of kHz QPOs to date. The lines correspond to the model prediction for different neutron star masses (the neutron star rotation has a small effect on these plots), where different points along the line correspond to different radii of emission. The model lines in Fig. 5.5 go though the cloud of points and the three qualitative predictions of the model are observed: positive correlation at low frequencies (the points on this branch come from the only source where low frequencies were observed, Circinus X-1 [Boutloukos *et al.* (2006)]), negative correlation at high frequencies, no ν_r value above 400 Hz.

A more direct look at the data is to plot one kHz QPO frequency versus the other (see Fig. 5.6) where different subclasses of

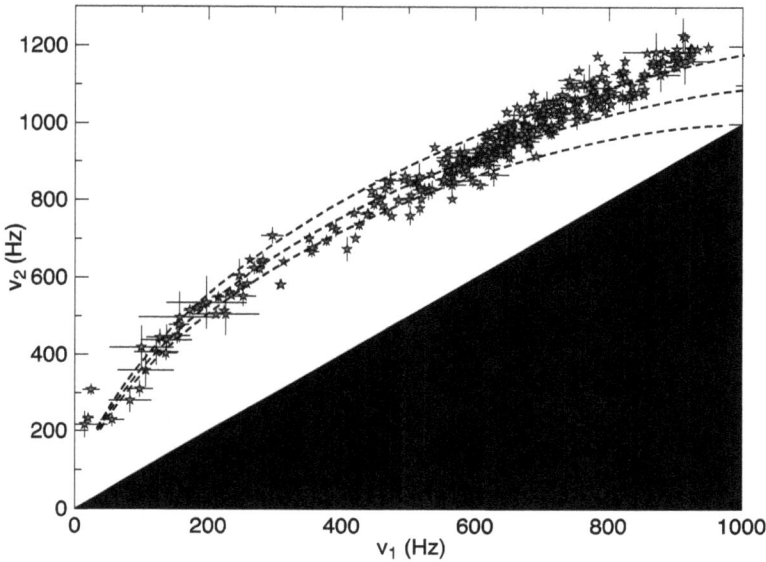

Fig. 5.6 Plot of one kHz QPO frequency versus the other. The dark area corresponds to a lower frequency higher than the upper one. The dashed lines correspond to the same model lines in Fig. 5.5.

NS binaries are shown. There are deviations at high frequencies, but the $2M_\odot$ model is a viable representation of the data.

In addition to the high-frequency features, the RPM interprets the low-frequency QPO seen in NS binaries and BHBs (also for NS there are more types of QPOs, but one of them can be associated to the one seen from BHBs) as ν_{nod}. In the weak field approximation, this frequency scales with the square of ν_ϕ (the Lense–Thirring effect). This dependence has been observed in NS binaries [Stella and Vietri (1998); Psaltis *et al.* (1999)].

One QPO in one source stands out: a peak in the range 35–50 Hz has been detected from a NS source that also showed a kHz QPO. These two features are inconsistent with being RPM frequencies originating from the same radius. More observations are needed to confirm the nature of the low-frequency QPO and the mismatch with the model, but this will have to wait for the source to be observable again, as it was a transient system.

Most of the attention has been on NS binaries, since as shown above the few detections of high-frequency features in BHBs did not correspond to detections of low-frequency QPOs. However, there is an observed correlation that suggests that high-frequency peaks could be observed also at low frequencies (associated to larger radii) and in that case they do not appear as peaks: the so-called PBK correlation [Psaltis, Belloni, and van der Klis (1999); Belloni, Psaltis, and van der Klis (2002)]. The correlation is shown in Fig. 5.7: this is the original correlation [Belloni, Psaltis, and van der Klis (2002)], but some incorrect points have been removed and two new points have been added. Here non-homogeneous quantities are plotted. For kHz QPOs, the low-frequency QPO frequency is plotted vs. that of the low-frequency QPO. For sources in the hard states (see PDS in

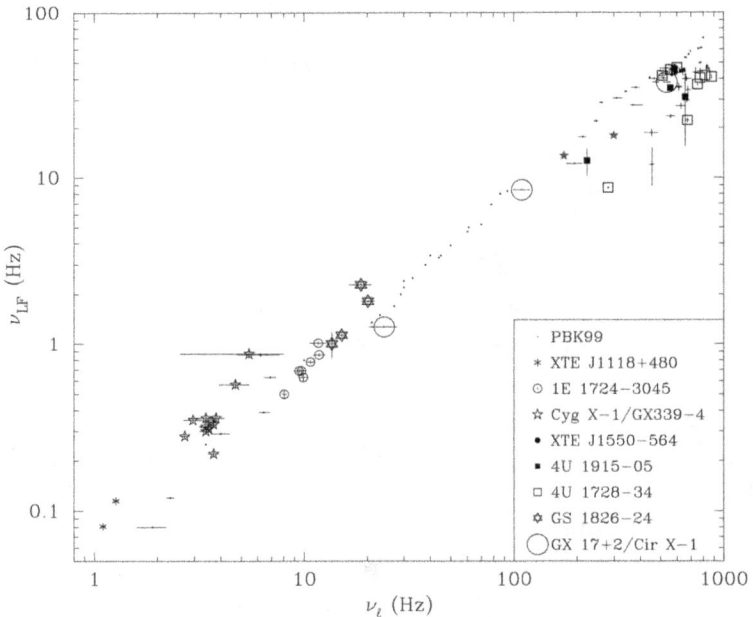

Fig. 5.7 The PBK correlation as published in its second version [Belloni, Psaltis, and van der Klis (2002)]. Some points have been removed as they are not significant and two points have been added, the stars in the upper right of the diagram, corresponding to the detections discussed below.

Fig. 5.3), both NS and BH, the characteristic frequency of the broad shoulder above it (see Fig. 5.3, where it is around 2 Hz) vs. the low-frequency QPO. In other words, while the x axis is always the frequency of the low-frequency QPO, the y axis for $x < 20$ Hz represents a broad component, while for $x > 100$ Hz it is a high-frequency peak. The points in between are from the same Circinus X-1 source that allowed to measure the left portion of Figs. 5.5 and 5.6. This correlation appears very tight and connects NS ad BH sources, as well as narrow and broad components, clearly suggesting a common physical mechanism. Indeed the RPM does interpret this correlation without needing additional parameters [Stella, Vietri, and Morsink (1999)].

It is interesting to note that this correlation, which covers three orders of magnitude, has been extended by two more orders of magnitude by including frequencies of dwarf-nova oscillations observed in the optical band from cataclysmic variables, binary systems containing white dwarfs [Warner, Woudt, and Pretorius (2003)]. While the correlation between the WD frequencies is good and it connects to the PBK, it is not clear how to link signals detected in different bands and from different systems, as the X-ray emitting region of an X-ray binary does not exist in a cataclysmic variable, due to the size of the star.

5.5.2. *The epicyclic resonance models*

As mentioned above, some of the observed pairs of high-frequency oscillations in BHBs have been observed to be close to a 3:2 ratio. This has led to the development of another class of local models, where the special radius associated to the QPOs is identified by resonances. In particular, there is a radius at which the radial ν_r and vertical ν_θ epicyclic frequencies have simple integer ratios (2:1, 3:2). This can lead to resonance [Kluźniak and Abramowicz (2001, 2002); Abramowicz and Kluźniak (2001)]. Additionally, also a resonance between ν_r and ν_ϕ has been considered [Török *et al.* (2005)]. The few available detections of QPO pairs in BHBs do not allow to check further the validity of the model, but the fact that the observed frequencies are observed only to show little variability is naturally explained within these models, since for a given object the resonance radii cannot change. Of course sharp changes from one resonance to the next one can in principle be observed, but no continuous shifts. Indeed, this model cannot be easily applied to neutron stars, where the frequencies vary and span a rather large range, without invoking an additional unknown mechanism that brings these changes, at which point one of the main predictions of the model does not exist anymore. Moreover, these models do not include low-frequency oscillations.

5.5.3.   *Other models*

Other models have been proposed and most of them are based on some of the fundamental frequencies of motion shown above. Both the RPM and the resonance models offer an interpretation for the observed centroid frequencies, but they do not include the presence of an accretion flow, nor they address the nature of the modulated emission. An extended approach to the study of the low-frequency QPOs in the Lense–Thirring hypothesis was presented a few years ago [Ingram, Done, and Fragile (2009); Ingram and Done (2011, 2012); Ingram and Motta (2014)]. To go beyond the single-radius test-particle approach, the model considers the precession of an extended hot region in the accretion flow, from which the observed QPO frequency would arise from a solid-body-like nodal precession that would depend on the outer radius of the region, located inside a truncated disk. The model allows the interpretation of observed correlations and a modulation of the iron line centroid in H 1743-322 has been observed to match its prediction due to variable irradiation of the accretion disk by the inner region (see Fig. 5.8) [Ingram *et al.* (2016, 2017)].

A model (AEI, accretion-ejection model) has been proposed based on a disk instability involving a spiral instability driven by magnetic stress from a poloidal field. The spiral extracts energy from the disk and transfers it to a corona, which then powers the ejection of relativistic jets [Tagger *et al.* (2004)]. The model has also been

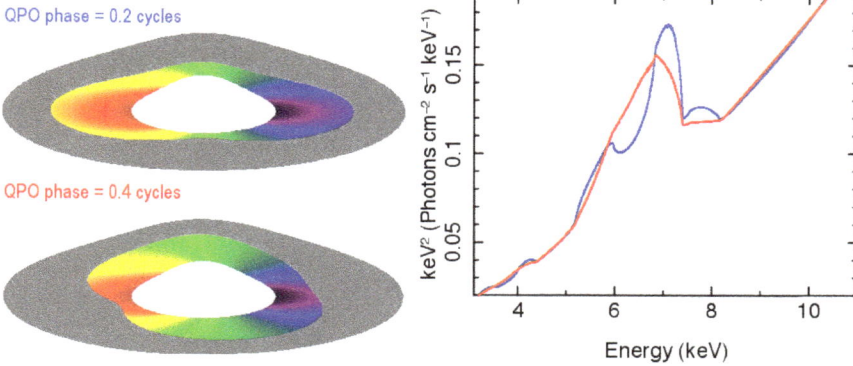

Fig. 5.8   Left: Ray-tracing representation of the illuminated part of the disk by the precessing inner hot flow at two different QPO phases for observations of H 1743-322. Right: Corresponding best fit line profiles [Ingram *et al.* (2017)].

extended to interpret high-frequency QPOs [Tagger and Varnière (2006)].

Global disk oscillation models involving different oscillation modes have been proposed and recently the first instances of QPO signals have been reproduced through hydrodynamical and MHD simulations. The discussion of these models goes beyond the scope of this article and can be found elsewhere [Belloni and Stella (2014)].

5.6. Where We Stand

It is clear that the existing data on high-frequency QPOs is so scarce that it is difficult to test the models above. Low-frequency QPOs have been widely observed, but their tendency not to be detected together with their high-frequency counterparts makes the testing of a full scenario problematic. However, out of the full archive of RossiXTE there exists one observation of a bright system, GRO J1655-40, where all three QPOs (one at low frequencies and two at high frequencies) have been detected (see Fig. 5.9) [Motta *et al.* (2014)].

With the equations above, under the assumption that the three frequencies correspond to GR frequencies at the same radius R, it was possible to estimate R and the mass and spin of the black hole with high accuracy: $R = 5.677 \pm 0.035\ r_g$, $M = 5.307 \pm 0.066\ M_\odot$,

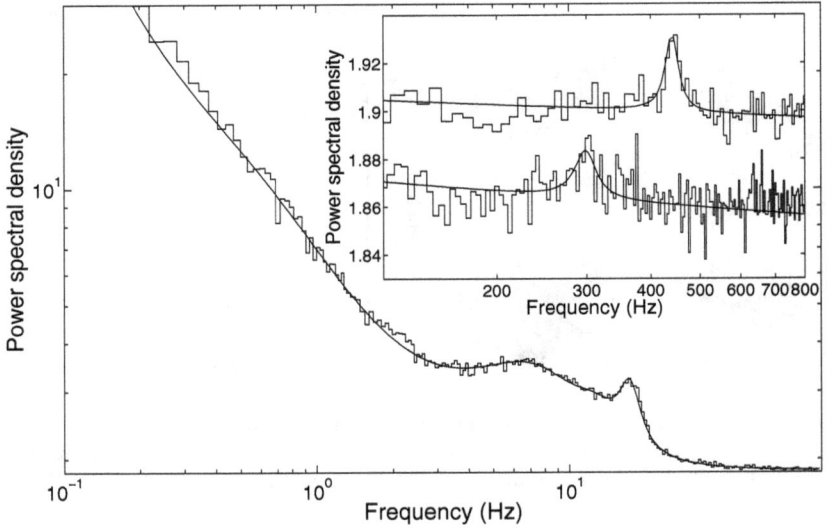

Fig. 5.9   PDS of the RossiXTE observation of GRO J1655-40 with three simultaneous oscillations. The main panel shows the 18 Hz low-frequency QPO, the inset shows the two high-frequency counterparts (in two different energy bands) [Motta *et al.* (2014)].

$a = 0.286 \pm 0.003$. Despite the small error bars, the derived mass matches the value obtained from dynamical measurements in the optical, $m = 5.4 \pm 0.4\,M_\odot$ [Motta *et al.* (2014); Beer and Podsiadlowski (2002)]. Since the only two parameters for the black hole are estimated, this also allows to derive a value for the ISCO $r_{ISCO} = 5.031 \pm 0.009\ r_g$. There are no other instances of three simultaneously detected QPOs, but there are many detections of the low-frequency QPO in GRO J1655-40 spanning the range 0.01–27 Hz. Since spin and mass are fixed, it was possible to test the hypothesis that the broad components observed together with low-frequency QPO represent relativistic frequencies (the PBK correlation, see above). Figure 5.10 shows that they fit very well the expected correlation with the low-frequency QPO [Motta *et al.* (2014)]. Notice that although the low-frequency QPOs span a very large range, they do not exceed the range of values expected at ISCO, which are not expected to be present. Moreover, the relative width of the three

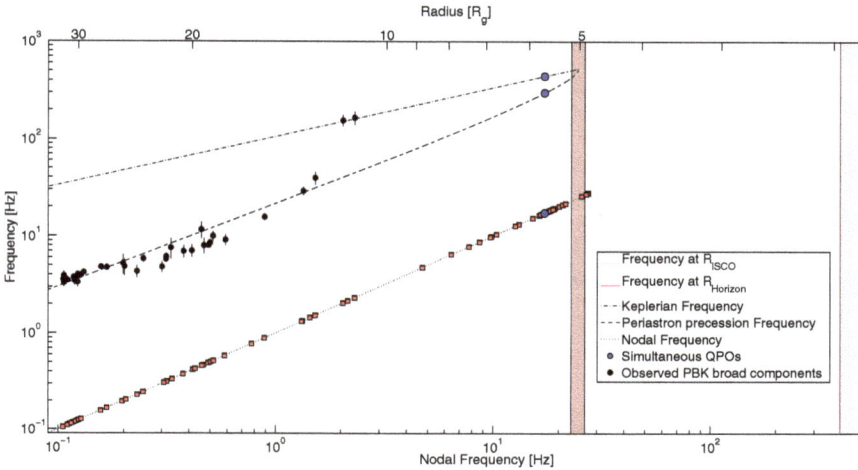

Fig. 5.10   Frequency correlation for the RossiXTE observations of GRO J1655-40 with QPOs, with the low-frequency in the abscissa [Motta *et al.* (2014)]. The red points mark low-frequency QPOs (therefore they mark a 1:1 line), the other points represent high-frequency QPOs and broad features (see text). The blue points are the three frequencies visible in Fig. 5.9. The dashed lines are the RPM predictions based on the three blue points, while the red band indicates the range of the expected maximum frequencies for the oscillations due to the presence of the ISCO. The upper x-axis shows the radii corresponding to the frequencies on the lower axis.

peaks can be explained with a modest jitter of the radius R, making the result even more robust.

As the estimate of the "correct" black-hole mass strengthens the association between observed and predicted frequencies, it was possible to apply the model to the second best observation, when another bright system (XTE J1550-564) was observed to show *two* simultaneous QPOs, one at low frequencies and one at high frequencies. With two frequencies the equations cannot be solved, but the inclusion of the optically measured black-hole mass leads to the estimate of $a = 0.34 \pm 0.01$ [Motta *et al.* (2014)]. Also in this case, the extension to lower frequencies through broad features is well interpreted by the model (see Fig. 5.11), as well as the relative width of the two peaks.

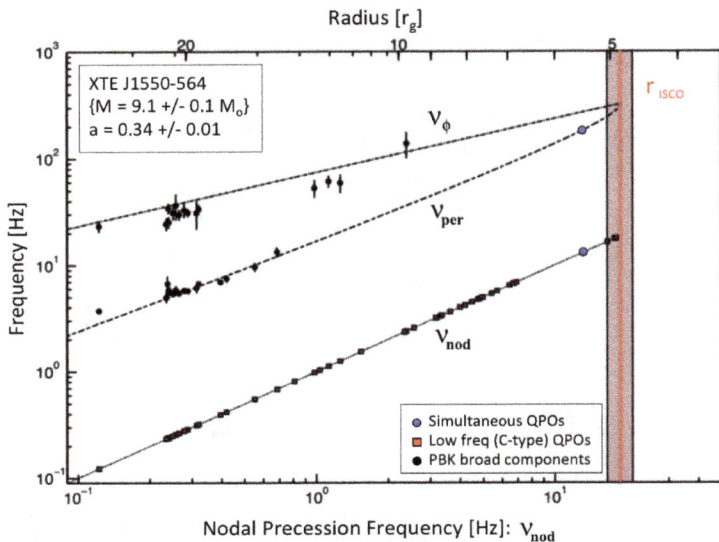

Fig. 5.11   Frequency correlation for the RossiXTE observations of GRO J1550-40, same plot as Fig. 5.10 [Motta *et al.* (2014)].

5.7. Conclusions

After years of observations of low- and high-frequency variability from BHBs we now have a number of models to test, but unfortunately too few detections of high-frequency QPOs to reach a firm conclusion. It is very likely that this is only a problem of sensitivity, as also shown by the fact that all detections were obtained when the sources were in their brightest stages. If the results shown in the previous section are confirmed (they are based on two observations only), the RPM model has led to the direct measurement of a stellar black-hole mass and therefore to the direct evidence of the presence of a black hole in the system, as well as to the existence of an ISCO and the confirmation of the relativistic precession frequencies in the strong-field regime.

What is needed now is more realistic models and new more sensitive observations. Models that take into account the presence of an accretion flow rather than simple test particles are being developed and have been shown above. What is still missing is a solid determination of the emission process that leads to the production of oscillations that can be as strong as 20% in rms. At the same time, new instruments are being developed. At the time of writing, the Indian X-ray satellite AstroSat is operative and is yielding data comparable in quality to those of RossiXTE. No new high-frequency oscillations have been reported yet, but observations are in place. For the future, larger missions are being proposed, such as eXTP [Zhang *et al.* (2017)], with instruments capable of collecting more photons and reaching a much higher sensitivity for fast variations.

References

Abramowicz, M.A. and Kluźniak, W., *Astron. Astrophys.* **374**, L19 (2001).
Arévalo, P. and Uttley, P., *Mon. Not. Roy. Astron. Soc.* **367**, 801 (2006).
Beer, M.E. and Podsiadlowski, P., *Mon. Not. Roy. Astron. Soc.* **331**, 351 (2002).
Belloni, T.M., arXiv:1803.03641 (2018).
Belloni, T.M. and Hasinger, G., *Astron. Astrophys.* **227**, L33 (1990).
Belloni, T.M., Psaltis, D., and van der Klis, M., *Astrophys. J.* **572**, 392 (2002).
Belloni, T.M. *et al.*, *Astron. Astrophys.* **440**, 207 (2008).
Belloni, T., Méndez, M., and Homan, J., *Mon. Not. Roy. Astron. Soc.* **376**, 1133 (2007).
Belloni, T.M., Sanna, A., and Méndez, M., *Mon. Not. Roy. Astron. Soc.* **426**, 1701 (2012).
Belloni, T.M. and Altamirano, D., *Mon. Not. Roy. Astron. Soc.* **432**, 10 (2013).
Belloni, T.M and Altamirano, D., *Mon. Not. Roy. Astron. Soc.* **432**, 19 (2013).
Belloni, T.M. and Stella, L., *Spa. Sci. Rev.* **183**, 43 (2014).
Belloni, T.M. and Motta, S.E., in *Astrophysics of Black Holes: From Fundamental Aspects to Latest Developments*, ed. C. Bambi (Springer-Verlag Berlin Heidelberg, 2016).
Boutloukos, S. *et al.*, *Astrophys. J.* **653**, 1435 (2006).
Casares, J. and Jonker, P.G., *Spa. Sci. Rev.* **183**, 223 (2014).
Davis, S.W. and Hubeny, I., *Astrophys. J. Suppl.* **164**, 530 (2006).
Fabian, A.C. and Ross, R.R., *Spa. Sci. Rev.* **157**, 167 (2010).
Kluźniak, W. and Abramowicz, M.A., arXiv:astro-ph/0105057 (2001).
Kluźniak, W. and Abramowicz, M.A., arXiv:astro-ph/0203314 (2002).
Kramer, M. *et al.*, *Science* **314**, 97 (2008).
Ingram, A., *Astron. Notes* **337**, 385 (2016).
Ingram, A., Done, C., and Fragile, P.C., *Mon. Not. Roy. Astron. Soc.* **397**, L101 (2009).
Ingram, A. and Done, C., *Mon. Not. Roy. Astron. Soc.* **415**, 2323 (2011).
Ingram, A. and Done, C., *Mon. Not. Roy. Astron. Soc.* **419**, 2369 (2012).
Ingram, A. and Motta, S., *Mon. Not. Roy. Astron. Soc.* **444**, 2065 (2014).

Ingram, A. *et al.*, *Mon. Not. Roy. Astron. Soc.* **464**, 2979 (2016).

Ingram, A. *et al.*, *Mon. Not. Roy. Astron. Soc.* **461**, 1967 (2017).

Lamb, F.K. *et al.*, *Nature* **317**, 681 (1985).

McClintock, J.E., Narayan, R., and Steiner, J.F., *Spa. Sci. Rev.* **183**, 295 (2014).

Miller, M.C. and Miller, J.M., *Phys. Rep.* **548**, 1 (2014).

Morgan, E.H., Remillard, R.A., and Greiner, J., *Astrophys. J.* **482**, 993 (1997).

Motta, S.E. *et al.*, *Mon. Not. Roy. Astron. Soc.* **437**, 2554 (2014).

Motta, S.E. *et al.*, *Mon. Not. Roy. Astron. Soc.* **439**, L65 (2014).

Nolan, P.L. *et al.*, *Astrophys. J.* **246**, 494 (1981).

Psaltis, D. *et al.*, *Astrophys. J.* **520**, 763 (1999).

Psaltis, D., Belloni, T., and van der Klis, M., *Astrophys. J.* **520**, 262 (1999).

Psaltis, D., *Living Reviews in Relativity* **11**, 9 (2008).

Ratti, E.M., Belloni, T.M., and Motta, S.E., *Mon. Not. Roy. Astron. Soc.* **423**, 694 (2012).

Ross, R.R. and Fabian, A.C., *Mon. Not. Roy. Astron. Soc.* **381**, 1697 (2007).

Shakura, N.I. and Sunyaev, R.A., *Astron. Astrophys.* **24**, 337 (1973).

Steiner, J.F. *et al.*, *Mon. Not. Roy. Astron. Soc.* **416**, 941 (2011).

Stella, L. and Vietri, M., *Astrophys. J.* **492**, L59 (1998).

Stella, L. and Vietri, M., *Phys. Rev. Lett.* **82**, 17 (1999).

Stella, L., Vietri, M., and Morsink, S.M., *Astrophys. J.* **524**, L63 (1999).

Tagger, M. *et al.*, *Astrophys. J.* **607**, 410 (2004).

Tagger, M. and Varnière, P., *Astrophys. J.* **652**, 1457 (2006).

Terrell, N.J., *Astrophys. J.* **174**, L35 (1972).

Török, G. *et al.*, *Astron. Astrophys.* **436**, 1 (2005).

van der Klis, M., in *Compact Stellar X-ray Sources*, eds. W. Lewin and M. van der Klis (Cambridge University Press, Cambridge, 2006).

Warner, B., Woudt, P.A., and Pretorius, M.L., *Mon. Not. Roy. Astron. Soc.* **344**, 1193 (2003).

Weisberg, J.M. and Huang, Y., *Astrophys. J.* **829**, 55 (2016).

Wijnands, R. and van der Klis, M., *Astrophys. J.* **522**, 965 (1999).

Zhang, S.N. *et al.*, *SPIE Proc.* **9905**, 99051Q (2017).

Chapter 6

Supermassive Black Holes in the Early Universe

José Antonio de Freitas Pacheco

Observatoire de la Côte d'Azur, Laboratoire Lagrange
06304 Nice Cedex, France
pacheco@oca.eu

The discovery of high redshift quasars represents a challenge to the origin of supermassive black holes. Here, two evolutionary scenarios are considered. The first one concerns massive black holes in the local universe, which in a large majority have been formed by the growth of seeds as their host galaxies are assembled in accordance with the hierarchical picture. In the second scenario, seeds with masses around 100–150 $M_\odot$ grow by accretion of gas forming a non-steady massive disk, whose existence is supported by the detection of huge amounts of gas and dust in high-z quasars. These models of non-steady self-gravitating disks explain quite well the observed luminosity–mass relation of quasars at high-z, indicating also that these objects do not radiate at the so-called Eddington limit.

6.1. Introduction

The cosmological nature of quasars (QSOs) was established in the early sixties [Schmidt (1963)]. An immediate consequence of the implied large distances for these objects was the realisation that QSOs were among the most powerful energy sources in the universe. Their luminosities are typically around 10^{46} erg·s^{-1} but the emission of some QSOs may exceed by one or two orders of magnitude that value. Edwin Salpeter was one of the first to propose that a super-massive black hole (SMBH) in a state of accretion could provide the necessary energy to explain the luminosities of QSOs [Salpeter (1964)]. If presently a large majority of the scientific community accepts that accreting SMBHs are the engines powering QSOs, a series of questions still remain to be answered. For instance, how are these SMBHs formed? If they grow by accretion what are the seeds and from where they come from? What is the gas accretion geometry: spherical or that of an inspiraling disk? In the last case, what are the viscous mechanisms responsible for the transfer of angular momentum?

In the local universe the presence of SMBHs in the center either of elliptical galaxies or bulges of spiral galaxies seems to be a well-established fact [Kormendy and Richstone (1995); Richstone *et al.* (1998); Kormendy and Gebhardt (2001)]. The black hole mass M is well correlated either with the stellar mass or the luminosity of the host bulge [Kormendy and Richstone (1995); Magorrian (1998); Marconi and Hunt (2003); Haring and Rix (2004); Graham (2007)] but, in particular, a tight correlation exists between the SMBH mass and the central projected stellar velocity dispersion σ [Ferrarese and Merrit (2000); Gebhardt *et al.* (2000); Merrit and Ferrarese (2001); Tremaine *et al.* (2002)]. The mechanism (or mechanisms) responsible for establishing the $M-\sigma$ relation is (are) not yet well determined but several scenarios have been put forward in the past years to explain the origin of such a relation. Self-regulated growth of black holes by feedback effects produced either by outflows or UV-radiation from QSOs, which affect also the star formation activity, is a possible mechanism able to reproduce the $M-\sigma$ relation [Silk and Rees (1998); Sazonov *et al.* (2005)]. A relation between these physical quantities can also be obtained from the picture developed by Burkert and Silk [Burkert and Silk (2001)] in which black holes grow at the expense of a viscous accretion disk and whose gas reservoir beyond the BH influence radius feeds also the formation of stars.

These investigations seem to point to a well-defined road leading to the formation of SMBHs: the growth of "seeds" by accretion inside the host galaxy. This picture is consistent with the fact that the present BH mass density agrees with the accreted (baryonic) mass density derived from the bolometric luminosity function of quasars [Soltan (1982); Small and Blandford (1992); Hopkins, Richards, and Hernquist (2007)] and with a negligible amount of accreted dark matter [Peirani and de Freitas Pacheco (2008)]. Seeds could be intermediate mass $(10^3–10^4)\,M_\odot$ black holes formed during the collapse of primordial gas clouds [Haehnelt and Rees (1993); Eisenstein and Loeb (1995); Koushiappas, Bullock, and Dekel (2004)] or during the core collapse of relativistic star clusters formed in starbursts, which may have occurred in the early evolution of galaxies [Shapiro (2004)]. Here, as it will be discussed later, seeds are assumed to be black holes with masses around 100–500 $M_\odot$ originated from the first generation of stars, supposed to be quite massive due to the absence of metals, which are the main contributors to the cooling of the gas.

The different correlations between the black hole mass and the dynamic or the photometric properties of the host galaxy suggest a gradual growth of the seed as the host galaxy itself is assembled. However this scenario seems to be inconsistent with the fact that up to now more than 40 bright QSOs have been discovered at high redshift (Wu *et al.*, 2015). The three QSOs having the highest redshift are J1061+3922 at $z = 6.61$, J1120+0641 at $z = 7.08$ and J1342+0928 at $z = 7.54$. The latter corresponds to an age of the universe of only 0.69 Gyr. Since most of these high redshift QSOs are associated with SMBHs having masses around $10^8–10^9\,M_\odot$, their growth was probably not gradual but rather fast in order to shine so early in the history of the universe. Thus, a possible issue is to admit the existence of two evolutionary paths leading to the formation of SMBHs: one in which seeds grow intermittently as their host galaxies are assembled and another in which seeds grow very fast in a timescale of less than 1 Gyr. These two possibilities will be discussed in the next sections of this article.

6.2. Spherical Accretion and the Eddington Limit

Many authors still consider in their investigations spherical accretion processes in which the mass inflow rate is controlled by the Eddington luminosity. In this case, it seems judicious to recall the physical assumptions that permit the derivation of the Eddington limit. The Euler equation describing a spherically symmetric inflow under the influence of gravitation and radiation pressure is

$$V\frac{dV}{dr} + \frac{1}{\rho}\left[\frac{d(P+P_r)}{dr}\right] + \frac{GM}{r^2} = 0. \tag{6.1}$$

In the above equation V is the radial flow velocity, P and P_r are respectively the gas and the radiation pressure, ρ is the gas density, G is the gravitational constant and M is the mass of the central object. The radial gradient due to the radiation pressure is given by

$$\frac{dP_r}{dr} = -\frac{1}{c}\int_0^\infty \kappa_\nu \phi_\nu d\nu = -\frac{\kappa}{c}\phi, \tag{6.2}$$

where κ is a suitable frequency average of the total absorption coefficient (including scattering) and ϕ is the total radiative flux. If the accreting gas envelope is highly ionized, the absorption of photons is essentially due to the Thomson scattering and, in this case

$$\kappa = \frac{\sigma_T}{\mu m_H}\rho, \tag{6.3}$$

where $\sigma_T = 6.65 \times 10^{-25}$ cm^2 is the Thomson cross-section, μ is the mean molecular weight and m_H is the proton mass. If the medium is optically thin and the radiation comes essentially from the deep inside region of the envelope, then the radiative flux in the outer regions is simply

$$\phi = \frac{L}{4\pi r^2}. \tag{6.4}$$

Combine Eqs. (6.2)–(6.4) and substitute into the Euler equation to obtain

$$V\frac{dV}{dr} + \frac{1}{\rho}\frac{dP}{dr} = -\frac{GM}{r^2} + \frac{\sigma_T L}{4\pi \mu m_H c r^2}. \tag{6.5}$$

Define now the Eddington luminosity as

$$L_E = \frac{4\pi GM\mu m_H c}{\sigma_T} = 1.76 \times 10^{38}\left(\frac{M}{M_\odot}\right) \text{erg}\cdot\text{s}^{-1}. \tag{6.6}$$

In this case, Eq. (6.5) can be rewritten as

$$V\frac{dV}{dr} + \frac{1}{\rho}\frac{dP}{dr} = -\frac{GM(1-\Gamma)}{r^2}, \tag{6.7}$$

where $\Gamma = L/L_E$.

Assume that the gas equation of state is given by $P = K\rho^\gamma$ and define also the adiabatic sound velocity as $a^2 = \gamma(P/\rho)$. Under these conditions, using the mass conservation equation to express the mass density gradient, after some algebra, Eq. (6.7) can be recast as

$$V\left(1 - \frac{a^2}{V^2}\right)\frac{dV}{dr} = \frac{2a^2}{r} - \frac{GM(1-\Gamma)}{r^2}. \tag{6.8}$$

The "critical" point of the flow in the usual mathematical sense corresponds to the point where both sides of Eq. (6.8) vanish. Hence, in order to have the continuity of the flow through the critical point, two conditions must be simultaneously satisfied, namely

$$V_* = a_*, \tag{6.9}$$

and

$$r_* = \frac{GM(1-\Gamma)}{2a_*^2}, \tag{6.10}$$

for the critical velocity and the critical radius respectively. These relations imply that the critical and the sonic points of the flow coincide (this is not always the case) and that the luminosity radiated from inside must be **smaller** than the Eddington value in order that the critical radius be real. Note that once the critical point is surpassed, the left side of Eq. (6.8) is negative, requiring imperatively that the right side be also negative or, equivalently, that $\Gamma < 1$. In other words, the spherical accretion of an optically thin envelope requires sub-Eddington conditions otherwise the inflow cannot be established. As we will see later, this requirement is weakened when the inflow geometry is modified as, for instance, in the case of an accretion disk.

As it was mentioned previously, many authors assume that the central black hole accretes mass with the envelope radiating near the Eddington limit. Since the Eddington luminosity is proportional to the mass of the black hole (see Eq. (6.6)), it results in an exponential growth with a timescale

$$\tau_E = \frac{\eta}{(1 - \eta)} \frac{c\sigma_T}{4\pi\mu m_H G} = 3.22 \times 10^8 \frac{\eta}{(1 - \eta)} \text{ yr}, \qquad (6.11)$$

where η is the accretion efficiency. Such a short timescale is often considered as an argument to explain the presence of SMBHs at high-z. However, as we have seen above, the Eddington limit is derived under conditions in which the envelope is optically thin and the opacity is due only to the Thomson scattering. The optical depth of the envelope is given by

$$\tau(r, \infty) = \int_r^\infty \frac{\sigma_T}{\mu m_H} \rho(r') dr' = 7.8 \times 10^{-6} \left(\frac{M}{M_\odot} \right) n_\infty, \qquad (6.12)$$

and the gravitational radius was taken as a lower limit in order to obtain the numerical value on the left side of the equation. For an optically thin envelope the condition $\tau < 1$ must be satisfied, imposing an upper bound to the black hole mass, namely

$$\frac{M}{M_\odot} < \frac{1.3 \times 10^5}{n_\infty}, \qquad (6.13)$$

where n_∞ is the gas particle density far from the influence radius of the black hole. Hence, if accreting spherically, SMBHs at high-z with masses around 10^8–$10^9 \, M_\odot$ will necessarily have an optically thick envelope and a different inflow regime. An optically thick envelope reduces the distance to the critical point and reduces also the accretion rate with respect to the optically thin case. In particular, when the radiation field is quite important, a second critical point may exist in the flow besides the hydrodynamical one, according to Nobili *et al.* [Nobili, Turolla, and Zampieri (1991)].

Accretion flows affected by radiation effects have been investigated by many authors in the past years [Maraschi, Reina, and Treves (1974); Flammang (1982); Milosavljevic, Couch, and Bromm (2009)]. The radiation from the accreting envelope is essentially due to the free-free emission. For the optically thin case, the resulting luminosity is proportional to the square of the accretion rate and inversely proportional to the BH mass, i.e., $L \sim \dot{M}^2/M$. The flow becomes nearly self-regulated when the optical depth of the infalling matter is greater than unity and under these conditions, the luminosity approaches the Eddington limit [Milosavljevic, Couch, and Bromm (2009)]. However, some authors claim that super-Eddington luminosities are possible if the black hole is embedded in a very dense gas cloud that decreases the importance of radiation pressure effects [Pacucci, Volonteri, and Ferrara (2015)]. Super-Eddington accretion rates were also found in some radiation-hydrodynamics simulations but based on one-dimensional geometry and particular conditions of the ambient gas [Inayoshi, Haiman, and Ostriker (2016)]. However, it is not certain whether such extreme conditions are sustainable considering the violent environments of the first galaxies where the medium is affected by the star formation activity and supernovae.

If the BH is moving with respect to the gas, the situation is rather different. After passing the BH, a conically shaped shock is produced in the flow in which the gas loses the momentum component perpendicular to the shock front. After compression in the shock, gas particles within a certain impact parameter will fall into the BH. One determinant factor describing the subsequent motion of the gas is the angular momentum. If the infalling gas has a specific angular momentum J that exceeds $2r_g c$, where $r_g = 2GM/c^2$ is the gravitational radius, the centrifugal forces will become important before the gas reaches the horizon. In this case, the gas will be thrown into near circular orbits and only after viscous stresses have transported away the excess of angular momentum will the gas cross the BH horizon (Shvartsman, 1971). In fact, the formation or not of a disk requires two conditions: the disk radius must be larger than the last stable circular orbit (equivalent to the condition $J > 2r_g c$) and must be smaller than the typical dimension of the shock cone, e.g., $l_s \approx 2r_g (c/u_\infty)^2$, where u_∞ is the BH velocity with respect to the gas. If the gas is highly turbulent, the velocity of eddies having a scale k is given roughly by

$$V_t \sim V_0 \left(\frac{k}{k_0} \right)^q. \tag{6.14}$$

In the case of a Kolmogorov spectrum, $q = 1/3$. However, it is more probable that the turbulent energy be dissipated mainly through shock waves and, in this case, the spectrum is steeper with $q \sim 1$ (Kaplan, 1954), a situation that will be assumed here. For our rough estimates, we will adopt typical values for the turbulence observed in our Galaxy, e.g., $V_0 \approx 10\,\text{km}\cdot\text{s}^{-1}$, $k_0 \approx 10\,\text{pc}$ (Kaplan and Pikel'ner, 1970). The specific angular momentum associated with eddies is $J \sim V_t k$ and the specific angular momentum of the accreted gas corresponds to eddies of the order of twice the scale of the capture impact parameter. Thus, the first condition for disk formation requires

$$M > 720 \left(\frac{u_\infty}{50\,\text{km}\cdot\text{s}^{-1}} \right)^4 M_\odot, \tag{6.15}$$

whereas the second requires

$$M < 3.5 \times 10^6 \left(\frac{u_\infty}{50\,\text{km}\cdot\text{s}^{-1}} \right)^3 M_\odot. \tag{6.16}$$

The conclusion of this brief analysis is that a disk can be formed after the shock front only if the BH mass is in the range 10^3 up to $10^6\,M_\odot$. Notice that these values depend strongly on the black hole velocity. Taking into account the restricted range of BH masses that allows the presence of a disk, the necessity of an adequate balance between the mass flow across the shock front and the flow throughout the disk that is controlled by viscous forces, as well as the variety of instabilities present in the flow after the shock. Under these conditions, the formation of a disk is rather uncertain. In this case, if a disk is not formed inside the accretion cone, the radiated luminosity is only a small fraction of the accretion power.

6.3. Intermittent Growth of Black Holes

As we have seen previously, the correlations between the black hole mass and the photometric or dynamical properties of the host galaxy suggest that the former grows as the latter is assembled. The different physical processes involved in the growth of black holes inside galaxies require a numerical treatment or, in other words, an appeal to cosmological simulations. In fact, there are different reasons justifying such an approach: first, because a significant volume of the universe can be probed; second, because the dynamics of dark matter and the hydrodynamics of the gas, including physical processes like heating, cooling, ionization of different elements can be taken into account self consistently. Moreover, it is possible to study environmental effects on the galaxies themselves as well as the chemical evolution of the interstellar and of the intergalactic gas, including the effects of supernovae and turbulent diffusion of heavy metals. The simulations permit to test star formation recipes, models for the growth of seeds and to investigate the influence of black holes on the environment when they are in an active phase.

6.3.1. *Cosmological simulations*

The results described in this section were all derived from simulations performed at the Observatoire de la Côte d'Azur (Nice) in these past years. Details of the code and some results can be found, for instance, in the papers by Filloux *et al.* (2010, 2011) or Durier and de Freitas Pacheco (2011). For the sake of completeness, a short summary of the main features of the code will be presented here.

All the simulations were performed in the context of the ΛCDM cosmology, using the parallel TreePM-SPH code GADGET-2 in a formulation, despite the use of fully adaptive smoothed particle hydrodynamics (SPH), that conserves energy and entropy (Springel, 2005). Initial conditions are established according to the algorithm COSMICS (Bertschinger, 1995) and the evolution of the structures are followed in the redshift range $60 \geq z \geq 0$. Ionization equilibrium taking into account collisional and radiative processes were included following Katz, Weinberg, and Hernquist (1998), as well as the contribution of the ionizing radiation background. The contribution of cooling processes such as collisional excitation of H I, He I and He II levels, radiative recombination, free-free emission and inverse Compton were also included, using the results by Sutherland and Dopita (1993). An interpolation procedure was adopted to take into account the enhancement of the cooling as the medium is enriched by metals. The cooling functions computed by those authors are adequate for highly ionized gases and for $T \geq 10^4$ K. At high redshifts $(100 > z > 20)$ and inside neutral gas clouds a residual electron fraction of about $n_e/n_H \approx 0.005$ is present (Peebles, 1993) which is enough to act as a catalyst in chemical reactions producing molecular hydrogen. H_2 cooling due to excitation of molecular rotational levels was introduced by using the results of Galli and Palla (1998). After the appearance of the first stars, the gas is enriched by trace elements like O, C, Si, Fe, responsible for a supplementary cooling mechanism. The UV background with $h\nu < 13.6$ eV is unable to ionize hydrogen (and oxygen) in neutral gas clouds but it can ionize carbon, silicon and iron, which are mostly singly ionized under these conditions. These ions have fine structure levels that can be excited by collisions either with electrons or atomic hydrogen, constituting an important cooling mechanism at low temperatures, which was included in the code. The UV radiation from young massive stars capable of ionizing the nearby gas was computed for different ionization species of hydrogen and helium, representing not only an additional (local) source of ionization but also of heating.

Feedback processes like the return of mass to the interstellar medium, supernova heating and chemical enrichment were all taken into account. The return of mass to the interstellar medium was computed by assuming that the initial mass function (IMF) of stars with metallicities $[Z/H] < -2.0$ is of the form $\zeta(m) \propto m^{-2}$ while stars more metal-rich are formed with a Salpeter IMF, e.g., $\zeta(m) \propto m^{-2.35}$. Stars in the mass range 40–80 $M_\odot$ leave a 10 $M_\odot$ black hole as a remnant, whereas a 1.4 $M_\odot$ neutron star is left if progenitors are in the mass range 9–40 $M_\odot$ or a white dwarf remnant otherwise. The mass lost by the "stellar particle" is redistributed according to the SPH kernel among the gas particle neighbors and velocities are adjusted in order to conserve the total momentum in the cell. Moreover, the removed gas (except that ejected by supernovae, as discussed below) keeps its original chemical composition, contributing to the chemical budget of the medium. In fact, AGB stars, planetary nebulae and WR stars enrich the medium in He, C, N, but these contributions are not taken into account in the present version of the code.

Supernova explosions are supposed to inject both thermal and mechanical energy in the interstellar medium. Past investigations have shown that, when heated, the nearby gas cools quite rapidly and the injected thermal energy is simply radiated away. However, when energy is injected in the form of kinetic energy, the star formation process is affected (Navarro and White, 1993). In the present simulations, supernova explosions were supposed to inject essentially mechanical energy into the interstellar medium through a "piston" mechanism, represented by the momentum carried by the ejected stellar envelope. The distance D_p covered by such a "piston", ejected with a typical velocity $V_{ej} \sim 3000 \, \text{km} \cdot \text{s}^{-1}$ in a time interval Δt is $V_{ej}\Delta t$. Under this assumption a "stellar cell" is defined including all gas particles inside a spherical volume of radius D_p that will be affected by the "piston". The released energy is redistributed non-uniformly among these particles. In this process, it is expected that the closest gas particles receive more energy than the farthest ones. This was achieved by assigning to each gas particle j a distance-dependent weight $w_j(r) = A/r_{ij}^n$, where r_{ij} is the distance between the "ith-stellar" particle (host of the SN explosion) to the jth gas particle inside the cell. The normalization constant is defined by $A = 1/\sum_j r_{ij}^n$. Different values of the exponent ($n = 2, 4$) were tested. Supernovae do not only contribute to the energy budget of the interstellar medium but also inject heavy metals, leading to a progressive chemical enrichment of galaxies as well as of their nearby environment. Such a progressive enrichment was treated by an adequate algorithm capable of simulating the turbulent diffusion of metals through the medium.

In the code, BHs are represented by collisionless particles that can grow in mass, according to specific rules that mimic accretion or merging with other BHs. Possible recoils due to a merging event and to the consequent emission of gravitational waves were neglected. BHs are assumed to merge if they come within a distance comparable to or less than the mean inter-particle separation. Seeds are supposed to have been formed from the first (very massive) stars and are supposed to have a mass of about 100 $M_\odot$. An auxiliary algorithm finds potential minima where seeds are inserted in the redshift interval

15–20. A fraction of the energy released during the accretion process is re-injected into the medium along two opposite "jets" aligned along the rotation axis of the disk, modeled by cones with an aperture angle of 20° and extending up to distances of about 300 kpc. The adopted expression for the power of the jets is essentially that given by the simulations of Koide *et al.* (2002).

6.3.2. *Properties of simulated SMBHs*

One of the main aspects of the growth of seeds by gas accretion concerns the fact that masses do not increase continuously. In the hierarchical model, galaxies are assembled in the filaments of the cosmic web or in the junction of filaments where clusters are formed. In filaments galaxies may capture fresh gas that will feed their central black holes. Gas may also come from merging events. However, from time to time the gas in the central region is exhausted and the process of growth stops until a new episode of gas capture is able to replenish the vicinity of the black hole. In fact, the amount of gas in the central regions of the host galaxy is controlled by the capture processes and internal processes like star formation and feedback from supernovae and the black hole itself. In Fig. 6.1 the individual growth of some

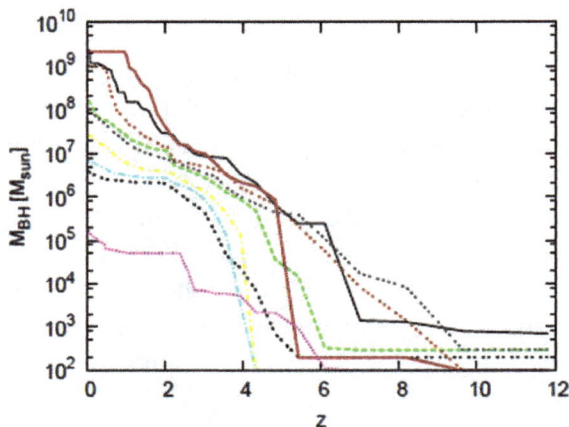

Fig. 6.1   Mass evolution of some individual black holes derived from simulations. Phases of activity or phases in which the black hole is in a dormant state can be clearly identified.

simulated black holes is shown. It is possible to verify that there are periods during which the black hole mass remains constant (case of a "dormant" black hole) and periods of gas accretion during which the black hole mass increases. In such a phase, the galaxy has an active nucleus, being associated to an AGN or to a QSO. Notice that only at $z \leq 4$ some black holes having masses greater than $10^7 \, M_\odot$ have appeared.

During the activity phase, the associated accretion disk is quite luminous and the luminosity depends essentially on the accretion rate. For a given redshift it is possible to compute the total luminosity due to all active black holes and, consequently, to estimate the comoving luminosity density. Such a luminosity density can be compared with observational data, permitting to test the robustness of the simulations. Figure 6.2 compares the luminosity density evolution derived from simulations with data by Hopkins *et al.* (2007). The agreement is quite satisfactory suggesting that the main physical aspects of the growth process are reasonably taken into account in the simulations.

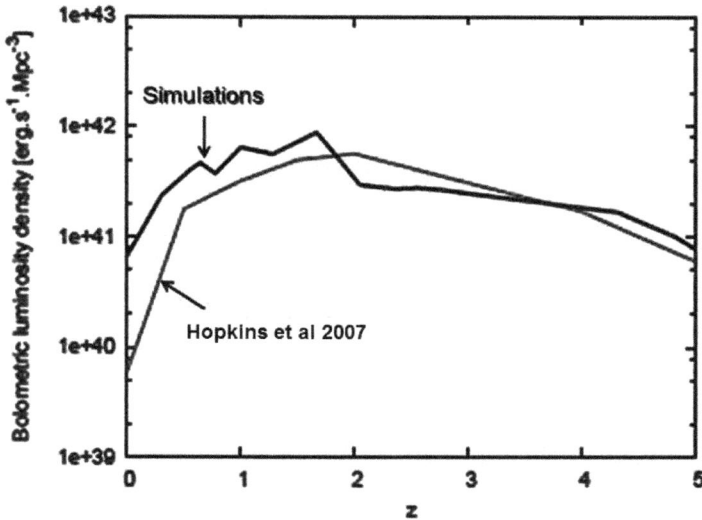

Fig. 6.2 Evolution of the luminosity density of active SMBHs derived from simulations compared with data on the luminosity density of QSOs.

Another successful comparison concerns the relation between the present SMBH mass with the central velocity dispersion of galaxies projected in the line of sight. This is done in Fig. 6.3. Black squares represent the masses of SMBHs at $z = 0$ derived from simulations while red squares represent data taken from the literature. There is a good agreement between simulated and observed data but some objects seem to have a higher black hole mass for the corresponding stellar velocity dispersion of their host galaxies. In particular, this is the case for NGC 5252 and Cygnus A as it can be seen in the figure. In our proposed scenario, these objects have not evolved intermittently but rather in a very short timescale in the early evolutionary phases of the universe, being presently the relics of such an active past.

If some SMBHs seem to have masses above that expected from the M–σ relation as Fig. 6.3 suggests, there are other arguments indicating that these objects followed indeed a different evolutionary path. At a given redshift, the simulations permit to compute the mass distribution of SMBHs. This is shown in Fig. 6.4, which indicates

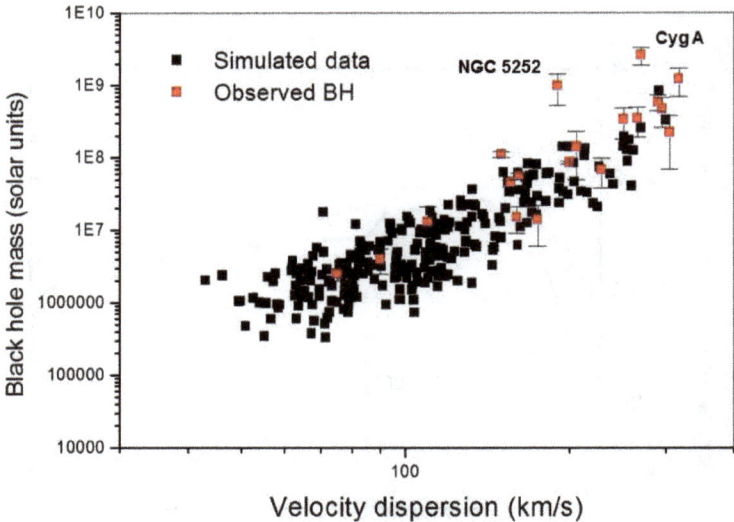

Fig. 6.3   Supermassive black hole mass versus projected central velocity dispersion of the host galaxy. Comparison between simulations and data. Identified objects have probably evolved along a different path.

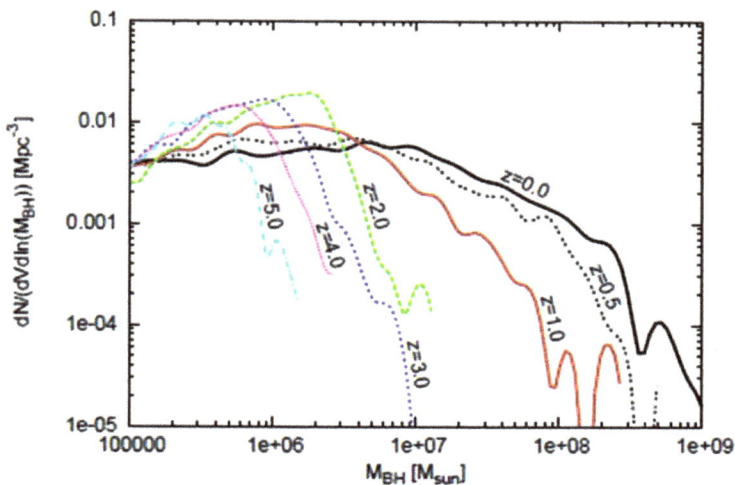

Fig. 6.4   Mass distribution of simulated black holes at different redshifts.

that no SMBHs with masses above $10^7 \, M_\odot$ is present at $z = 5$, in agreement with the evolution of individual black holes shown in Fig. 6.1. This means that the evolutionary path in which the BH mass grows as the host galaxy is assembled, which is probably the origin of the M–σ relation, is unable to explain the existence of very massive BHs in the early universe or SMBHs present today in bright galaxies like NGC 5252 or Cygnus A. In the next sections an alternative evolutionary path will be examined.

6.4. The Early Formation of SMBHs

As it was shown in the previous section, the coeval evolution of seeds and host galaxies is not able to explain the existence of bright QSOs at high redshift and the fact that in the local universe some objects have masses higher than that expected from the simulated M–σ relation.

Could a unique accretion disk form a SMBH in a timescale of about 1 Gyr? The answer to this question implies the solution of different related problems. The disk must be quite massive in order to provide enough gas to form a $10^9 \, M_\odot$ black hole and the angular momentum transfer must be very efficient in order to maintain a high accretion rate necessary to provide the observed luminosities as well as a short timescale for the growth of the seed. In fact, numerical simulations suggest that after the merger of two galaxies, a considerable amount of gas is settled into the central region of the resulting object. The gas loses angular momentum in a timescale comparable to the dynamical timescale (Mihos and Hernquist, 1996; Barnes, 2002), forming circumnuclear self-gravitating disks having masses in the range 10^6–$10^9 \, M_\odot$ and dimensions of about 100–500 pc.

Massive accretion disks are, in general, self-gravitating in their early evolutionary phases, a situation that affects the usual dynamics of disks controlled only by the gravitational forces due to the central body. In fact, models of non-steady self-gravitating accretion disks were computed by Montesinos and de Freitas Pacheco (2011, hereafter MP11) satisfying the aforementioned requirements or, in other words, they are luminous enough and they permit the growth of seeds in a short timescale, consistent with observations of bright QSOs at high z. Some aspects of the work by those authors will be shortly reviewed below, followed by the presentation of new results.

As mentioned before, the very early formation of a massive disk in the central region of a galaxy requires the presence of a large amount of gas. In fact, infrared sky surveys have discovered huge amounts of molecular gas (CO) at intermediate and high-z QSOs (Downes *et al.*, 1999; Bertoldi *et al.*, 2003; Weiss *et al.*, 2007). In particular, the detection of CO emission in the quasar J1148+5251 at $z = 6.42$ permitted an estimation of the molecular hydrogen mass present in the central region of the host galaxy that amounts to $M(H_2) \sim 10^{10} \, M_\odot$ (Walter *et al.*, 2009). Moreover, at least in the case of the quasar J1319+0950 ($z = 6.13$) there is a robust evidence that the gas is rotating (Shao *et al.*, 2017), suggesting the presence of a gaseous disk. More recently, observations of J1342+0928 ($z = 7.54$) indicate important amounts of gas and dust revealed by the infrared continuum and by the [C II] line emission (Venemans *et al.*, 2017). All these observations support the idea that some massive galaxies in their early evolutionary phases had large amounts of gas in their central regions, which could have formed the massive accretion disks required by our model.

If observations seem to support the scenario in which massive disks fed the seeds of SMBHs, the other question concerns the accretion timescale defining the growth of those seeds. The accretion rate is fixed by the mechanism of angular momentum transfer and depends on the gas viscosity mechanism. Presently there is no adequate physical theory able to describe the gas viscosity in the presence of turbulent flows or in the presence of magnetic fields. The angular momentum transfer is generally described by the formalism introduced almost forty five years ago by Shakura and Sunyaev (1973), in which the viscosity is due to subsonic turbulence and is parameterized by the relation $\eta = \alpha H c_s$, where η is the viscosity, $\alpha \leq 1$ is a free parameter of the theory, c_s is the sound velocity and H is the vertical scale of height of the disk. H is supposed to be of the same order as the typical (isotropic) turbulence scale. However, disks based on such a formalism are, in general, thermally unstable as demonstrated long time ago by Piran (1978).

It is well known that self-gravitating disks may be also unstable but, in some cases, such an instability can be a source of turbulence in the flow (Duschel and Britsch, 2006). Simulations of the gas inflow in the central regions of galaxies induced by the gravitational potential either of the stellar nucleus or the SMBH, reveal the appearance of highly supersonic turbulence with velocities of the order of the virial value (Regan and Haehnelt, 2009; Levine *et al.*, 2008; Wise, Turk, and Abel, 2008). No fragmentation is observed in such a gas despite of being isothermal and gravitationally unstable. This behavior can be explained if an efficient angular momentum transfer suppresses fragmentation. On the contrary, if the angular momentum transfer is inefficient, the turbulence decays and triggers global instabilities which regenerate a turbulent flow. Thus, one could expect that the flow will be self-regulated by such a mechanism. In this case, the flow must be characterized by a critical Reynolds number $\mathcal{R}$, determined by the viscosity below which the flow becomes unstable (de Freitas

Pacheco & Steiner 1973). This critical viscosity is given by

$$\eta = \frac{2\pi r V_\phi}{\mathcal{R}}, \tag{6.17}$$

where r is the radial coordinate and $V_\phi = r\Omega$ is the azimuthal velocity of the flow.

Another difference between "α"-disk and "critical viscosity" models is the local balance of energy that is fixed by the equilibration of the rate of the dissipated turbulent energy with the radiated and advected energy rates or, in other words

$$T_{r\phi} \frac{d\Omega}{dlgr} = \nabla \cdot F_{rad} + \varepsilon_{adv}. \tag{6.18}$$

In the above equation the left side represents the rate of turbulent energy dissipated per unit volume, the first term on the right side gives the rate per unit volume of radiated energy and finally, the last term represents the rate per unit volume of advected energy. In Eq. (6.18), $T_{r\phi}$ is the (r, ϕ) component of the stress tensor, Ω is the angular flow velocity, F_{rad} is the radiative flux and ε_{adv} is the rate per unit volume of advected energy. In the "α"-disk model the considered stress component is given by

$$T_{r\phi} = \alpha\rho H c_s \left(\frac{d\Omega}{dlgr} \right), \tag{6.19}$$

while in the "critical viscosity" model the stress is given by

$$T_{r\phi} = \frac{2\pi}{\mathcal{R}} \rho r^2 \Omega \left(\frac{d\Omega}{dlgr} \right). \tag{6.20}$$

The difference between the heating rates in both models implies that the temperature distribution along the disk is not the same and that the expected radiated spectrum of each disk model is also different as it will be discussed later.

In non-steady self-gravitating disk models, the dynamics of the disk evolves because the mass distribution changes with time as well as the mass of the central black hole. Near the BH the velocity is approximately Keplerian while beyond the transition region, where self-gravitation dominates, the rotational velocity decreases with distance more slowly than $1/\sqrt{r}$. Along the vertical axis, the disk is supposed to be in hydrostatic equilibrium. The vertical scale of height varies as the disk evolves. In early phases the disk is geometrically thick in the central regions due to radiation pressure effects. In late phases, the vertical scale of height increases with distance, approaching the behavior displayed by canonical non-self-gravitating models. Additional details, including a description of the numerical code used to solve the hydrodynamic equations, can be found in MF11 as mentioned previously.

Figure 6.5, adapted from MF11, shows some examples of models characterized by different values of the seed (100 or 1500 $M_\odot$) and of the critical Reynolds number (500, 1000 or 1500). Notice that an initial black hole of 100 $M_\odot$ can grow up to 3×10^8 $M_\odot$ in a timescale of only 10^8 yr if the critical Reynolds number is 500 (black curve). Inspection of Fig. 6.5 shows that for the same initial mass, if the Reynolds number is increased (red and green curves), the rate of

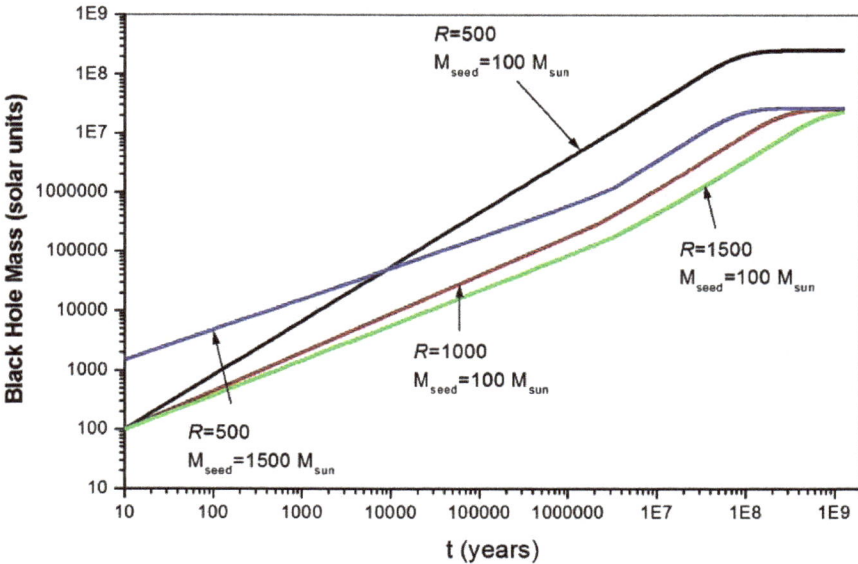

Fig. 6.5 Mass evolution of supermassive black holes for different values of the seed and the critical Reynolds number.

growth decreases. This can be explained by the fact that the accretion rate is inversely proportional to the viscous timescale, namely, $t_{vis}^{-1} \approx \eta/r^2 \approx \Omega/\mathcal{R}$. This is an immediate consequence of the fact that increasing the Reynolds number makes it more difficult to generate the turbulence. Therefore the angular momentum transfer is less efficient, reducing the accretion rate.

It is important to emphasize that while in the internal parts of the disk the gas flows inwards, the outer parts expand as a consequence of the transfer of angular momentum from inside to outside. Hence, only about 50% of the initial mass of the disk is in fact accreted by the black hole. In the region where the sign of the radial velocity changes (the "stagnation" point) a torus-like structure is formed, supporting the scenario of the so-called "unified model" of AGNs. The models by MF11 indicate that a substantial fraction of the expanding gas remains neutral with a temperature in the range 100–2000 K most of the time. In the case of our own galaxy, such a behavior could be related to the molecular ring of 2 pc radius observed around Sgr A* (Gusten, R. *et al.*, 1987). The physical conditions prevailing in the outskirts of the disk are favorable to star formation and could be an explanation for the presence of massive early-type stars located in two rotating thin disks detected in the central region of the Milky Way (Genzel *et al.*, 2003; Paumard *et al.*, 2006).

6.4.1. *Further tests of the model*

In the very beginning of the disk evolution, the accretion rate (and the luminosity) increases very rapidly and then remains more or less constant during most of the growth process. In the final phases, the accretion rate decays very fast once half of the disk mass is captured by the central black hole. Such a behavior can be seen in Figs. 6.2 and 6.3 of MF11. Depending on the initial disk mass and on the critical Reynolds number, the activity phase corresponding to the luminosity maximum lasts for about 2×10^7 up to 3×10^8 years.

Despite the fact that the accretion rate (and the luminosity) varies very little during the active phase, the spectral distribution of the radiation emitted by the disk evolves. Such a spectral evolution is due to time variations of the optical depth radial profile as well as time variations of the radial temperature distribution, as mentioned earlier. There is a continuous shift of the emission maximum toward longer wavelengths that is a consequence of the decreasing average disk temperature as a function of time. In general for wavelengths $\lambda \geq 0.15\,\mu$m the spectral intensity can be well represented by a power law, that is, $I_\lambda \propto \lambda^{-\alpha}$, where the power index is in the range $0.9 < \alpha < 1.3$, in agreement with values derived from most of quasar spectra.

The modeling of the spectral emission of the disk permits an estimate of the bolometric correction. Usually, the luminosity in a given wavelength is derived from observations of monochromatic fluxes and luminosity distances, which depend on the redshift. The bolometric luminosity can be computed by adopting an adequate correction. Nemmen and Brotherton (2010) have estimated the bolometric correction for luminosities at $\lambda = 0.30\,\mu$m based on models by Hubeny *et al.* (2000). The grid of models by the latter authors assimilates to each annulus of the disk an effective temperature and gravity, which are used to compute the emergent spectrum of an equivalent stellar atmosphere. The sum of the radiation from all annuli gives the resulting spectrum of the disk. However the effective temperature formula adopted by those authors is only adequate for a steady disk whose dynamics is dominated by the central black hole. In the case of non-steady self-gravitating disks the situation is rather different because both the local gravity and the effective temperature vary with time. Fortunately, the bolometric corrections for the monochromatic luminosities at $\lambda = 0.30\,\mu$m or at $\lambda = 3.6\mu$m do not vary too much during the active phase and a suitable average correction can be defined.

The adopted procedure to estimate such a bolometric correction requires an adequate choice of the representative parameters of the disk, since the seeds must be able to grow in timescales less than $1\,$Gyr and form SMBHs with masses larger than $5 \times 10^8\,M_\odot$. Figure 6.6 shows the surface "M–age–$\mathcal{R}$" derived from a grid of

Reynolds Number

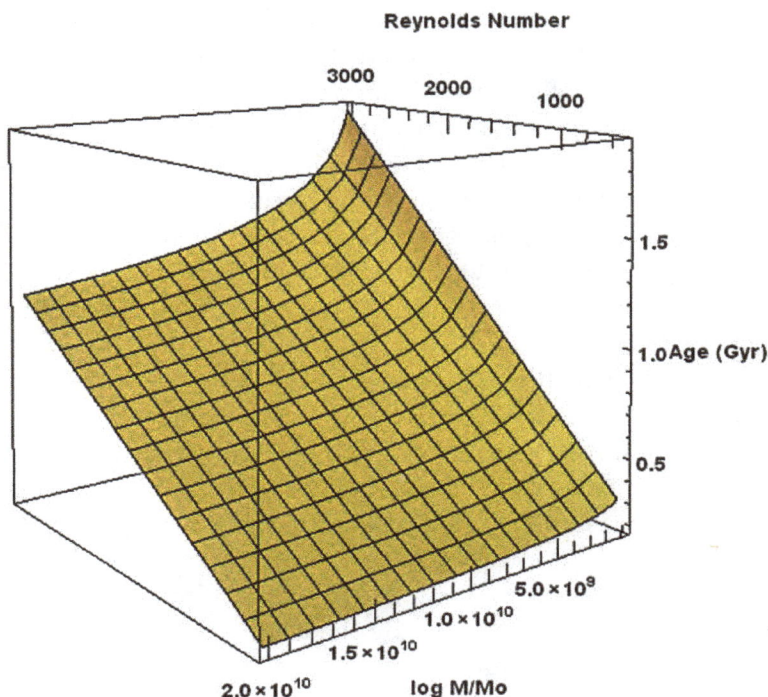

Fig. 6.6 Surface defining the space of parameters: mass of the black hole, age and critical Reynolds number.

models where the seed mass was fixed to $100\,M_\odot$. It is worth mentioning that in such a plot the parameter "age" means the timescale required for the seed to accrete 50% of the initial disk mass, the same definition that was adopted by MF11. Inspection of Fig. 6.6 indicates that Reynolds numbers in the range $1000 < \mathcal{R} < 2500$ are required in order to satisfy those constraints. Then, bolometric corrections at $\lambda = 0.30\,\mu$m and at $\lambda = 3.6\,\mu$m were computed for a series of models characterised by $\mathcal{R} = 2200$, mass of the seed equal to $100\,M_\odot$ and different initial disk masses, corresponding to about twice the final black hole masses. After averaging the results from different models, the corrections are simply given by

$$\log L_{bol} = \log \lambda L_\lambda + 0.83 \quad \text{for } \lambda = 0.30\,\mu\text{m}, \tag{6.21}$$

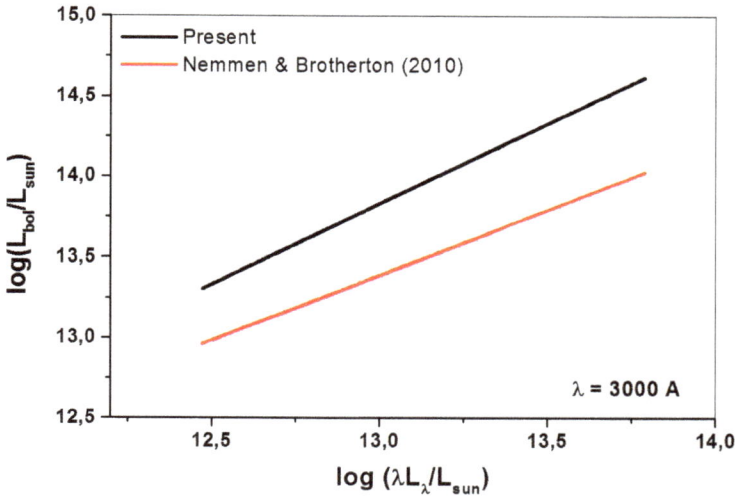

Fig. 6.7 Comparison between bolometric corrections based on steady and non-steady accretion disks.

and

$$\log L_{bol} = \log \lambda L_\lambda + 0.92 \quad \text{for } \lambda = 3.6\,\mu\text{m}, \qquad (6.22)$$

where the luminosities are given in erg·s^{-1}. In Fig. 6.7 the bolometric correction for $\lambda = 0.30\,\mu$m derived from these models is compared with the correction adopted by Nemmen and Brotherton (2010) based on steady and non-self-gravitating disk models. It should be emphasized that the present bolometric luminosities derived either from UV or infrared monochromatic luminosities are in very good agreement when the corrections above are applied.

The present disk model can also be tested by the comparison between theoretical predictions and data in the diagram of bolometric luminosity versus black hole mass. Data on high redshift QSOs ($z \geq 6.0$), including masses, monochromatic luminosities at $\lambda = 0.30$ or $3.6\,\mu$m and redshift, compiled by Trakhtenbrot *et al.* (2017) were used in the calculations. Black hole masses were estimated from the width of Mg II lines and monochromatic UV-luminosities were derived from the best-fit model of the Mg II emission line complex. Then, using the derived bolometric corrections, the bolometric

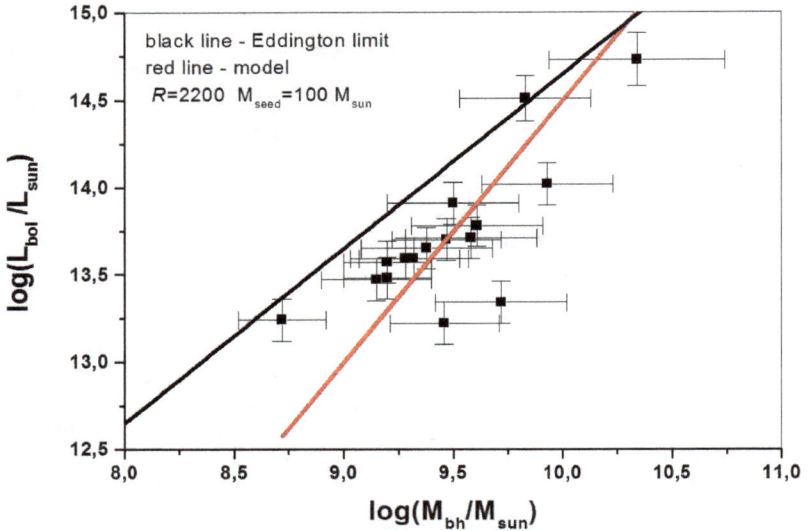

Fig. 6.8   Bolometric luminosity plotted against the black hole mass for objects with $z \geq 6.0$. The bolometric correction at $0.30\,\mu$m is given in the text. The red line corresponds to Eq. (6.23) with $\mathcal{R} = 2200$ and $M_{seed} = 100\,M_\odot$. The black line corresponds to the Eddington limit.

luminosity of each object was estimated and plotted as a function of the black hole mass in Fig. 6.8. The "mass–luminosity" relation derived from our models can be adequately represented by the fit

$$\frac{L_{bol}}{L_\odot} = 1.41 \left(\frac{500}{\mathcal{R}}\right) \left(\frac{M_{seed}}{100 M_\odot}\right)^{0.52} \left(\frac{M}{M_\odot}\right)^{1.5}, \qquad (6.23)$$

which depends essentially on the seed mass and on the critical Reynolds number. Such a relation for $M_{seed} = 100\,M_\odot$ and $\mathcal{R} = 2200$ is shown in Fig. 6.8 as a red line.

In Fig. 6.8 the expected relation for the Eddington limited luminosity is also shown (see Eq. (6.6)), frequently used to estimate the mass of the black hole. Notice that the theoretical "M–L" relation approaches the Eddington limit for black hole masses greater than $10^{10}\,M_\odot$. It is worth mentioning that identifying the bolometric luminosity with the Eddington limit leads to an underestimate of the black hole mass, as it can be seen in the plot, where the majority of the data points are below the expected Eddington limit line. On the other hand, the theoretical "*L–M*" relation displayed in Fig. 6.8 shows clearly that our disk models radiate below the Eddington limit.

It should be emphasized that in the case of accretion disks, the balance between gravity and radiation pressure along the vertical axis must be considered locally. The disk is locally stable if the radiative flux along the vertical direction is not greater than a critical flux limit given by

$$F_{rad} = \frac{\sqrt{3}}{4\pi} \left(\frac{m_H c}{\sigma_T} \right) \left(\frac{\tau_s}{\tau_{ff}} \right)^{1/2} g_z. \tag{6.24}$$

In the above equation τ_s is the optical depth due to electron scattering, τ_{ff} is the optical depth due to free-free absorption and g_z is the local vertical gravitational acceleration. The condition expressed by Eq. (6.24) is valid in the inner regions of the disk where the electron scattering dominates over free-free absorption and where radiation pressure effects are more important. When the vertical radiation flux is higher than the critical value, the hydrostatic equilibrium is destroyed and outflows can be generated. Three-dimensional radiation magnetohydrodynamical simulations were performed by Jiang *et al.* (2017), who have studied the evolution of an accretion disk with torus centered on a $5 \times 10^8 \, M_\odot$ black hole. The radiation pressure in the internal regions of the disk may reach values up to 10^6 times the gas pressure under certain conditions, producing outflows. In these simulations, the angular momentum transfer is controlled by magnetohydrodynamic turbulence that is not the case of our models.

The present accretion disk models can be also tested in the diagram of black hole mass versus age (Fig. 6.9). This plot is simply the projection of the surface displayed in Fig. 6.6 on the considered plane "*M*–age". Since the age parameter defined above is not directly accessible from observations, the age of the universe derived from the observed redshift of the QSO was used to plot the objects listed by Trakhtenbrot *et al.* (2017). The age of the universe represents a robust upper limit to the age parameter of the model. Theoretical predictions shown in Fig. 6.9 (solid lines) were computed for the same value of the seed mass ($M_{seed} = 150 \, M_\odot$) and for two different

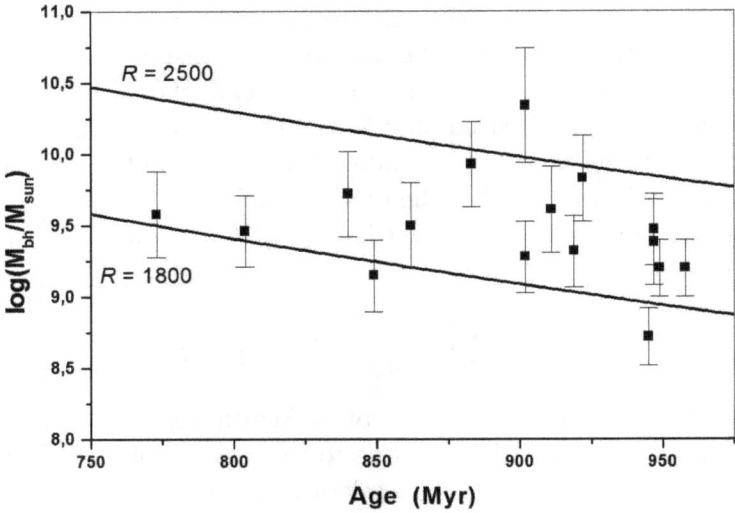

Fig. 6.9   Black hole masses versus age of the universe. Solid lines represent theoretical predictions from non-steady self-gravitating accretion disks.

critical Reynolds numbers: $\mathcal{R} = 1800$ and $\mathcal{R} = 2500$. These two values enclose in the plot most of the observed high-z QSOs, strongly constraining this fundamental parameter of the model.

6.5. Conclusions

Present astronomical data are not in contradiction with a scenario in which two different evolutionary paths exist for the formation of SMBHs from small mass seeds.

In the first evolutionary path, seeds having masses around $100\,M_\odot$ grow intermittently following the gradual assembly of the host galaxy, according to the hierarchical picture. In this case, the coeval evolution between the host galaxy and the seed must be investigated by cosmological simulations. This procedure is justified by the complexity of the physical mechanisms involved in the process of growth. As we have previously seen, these numerical experiments are able to reproduce the observed luminosity density of QSOs and the observed correlations between the black hole mass at $z = 0$ and the properties of the host galaxy like the stellar luminosity or the central projected stellar velocity dispersion. Despite these successful results, these simulations are unable to form SMBHs with masses around $10^9 \, M_\odot$ at high redshift, unless the masses of the seeds are dramatically increased to 10^5–$10^6 \, M_\odot$. Although this could be a possibility and despite some studies in this direction, such an alternative seems to be unrealistic.

The existence of bright QSOs at $z \approx 6$–7 and the difficulty for cosmological simulations to form these objects point toward a new direction, that is, the possibility of a very fast growth of seeds fed by massive accretion disks. This picture is supported by observations of large amounts of gas and dust in QSOs at high-z as discussed before. Models of non-steady self-gravitating disks in which the angular momentum transfer is controlled by turbulent viscosity were developed by Montesinos and de Freitas Pacheco (2011). These models have demonstrated that seeds can grow in timescales of the order of 1 Gyr or even less, being able to explain the main features of QSOs observed at high-z.

Further investigations on these "critical-viscosity" disks permitted an estimation of the bolometric correction that should be applied to monochromatic luminosities measured at $0.30 \, \mu\text{m}$ and $3.6 \, \mu\text{m}$. These corrections permitted a comparison of existing data with theoretical predictions in the luminosity–mass diagram. Such a plot strongly suggests that accretion disks radiate below the so-called Eddington limit. This means that black hole masses derived from such a limit are underestimated. Another useful diagram permitting the comparison of model predictions with data is the mass–age plot. Here it is necessary to recall the previous remarks, that is, the age derived from the redshift is the age of the universe at that moment, representing only a robust upper limit to the disk age. Nevertheless, despite such limitations, both diagrams permit to constrain the two important parameters of the model, namely, the mass of the seeds and the critical Reynolds number. The former is probably in the range 100–150 $M_\odot$ while the latter should be in the interval $1800 < \mathcal{R} < 2500$.

Finally, it is worth mentioning that some SMBHs in the local universe ($z = 0$) have masses above that expected from simulations.

This is the case of NGC 5252 and Cygnus A as already mentioned, but may also be the case of NGC 3115 and probably NGC 4594. This last object is more uncertain since its estimated mass is only 4.4 times greater than that expected from the simulated "$M\text{--}\sigma$" relation. These objects are probably the remnants of a fast growth occurred in the early evolutionary phases of the universe and not the consequence of a coeval evolution involving the seed and the host galaxy.

References

Barnes, J.E., *Mon. Not. Roy. Astron. Soc.* **333**, 481 (2002).

Bertoldi, F. *et al.*, *Astron. Astrophys.* **409**, L47 (2003).

Bertschinger, E., arXiv:astro-ph/9506070 (1995).

Burkert, A. and Silk, J., *Astrophys. J.* **554**, L151 (2001).

de Freitas Pacheco, J.A. and Steiner, J.E., *Astrophys. Sp. Sci.* **39**, 487 (1976).

Durier, F. and de Freitas Pacheco, J.A., *Int. J. Mod. Phys. E* **20**, 44 (2011).

Downes, D. *et al.*, *Astrophys. J.* **513**, L1 (1999).

Duschel, W.J. and Britsch, M., *Astrophys. J.* **653**, L92 (2006).

Eisenstein, D.J. and Loeb, A., *Astrophys. J.* **443**, 11 (1995).

Ferrarese, L. and Merrit, D., *Astrophys. J.* **539**, L9 (2000)

Filloux, Ch. *et al.*, *Int. J. Mod. Phys. D* **19**, 1233 (2010).

Filloux, Ch. *et al.*, *Int. J. Mod. Phys. D* **20**, 2399 (2011).

Flammang, R.A., *Mon. Not. Roy. Astron. Soc.* **199**, 833 (1982).

Galli, D. and Palla, F., *Astron. Astrophys.* **335**, 403 (1998).

Gebhardt, K. *et al.*, *Astrophys. J.* **539**, L13 (2000).

Genzel, R. *et al.*, *Astrophys. J.* **594**, 813 (2003).

Graham, A.W., *Mon. Not. Roy. Astron. Soc.* **379**, 711 (2007).

Gusten, R. *et al.*, *Astrophys. J.* **318**, 124 (1987).

Haehnelt, M.G. and Rees, M.J., *Mon. Not. Roy. Astron. Soc.* **263**, 168 (1993).

Haring, N. and Rix, H.-W., *Astrophys. J.* **604**, L89 (2004).

Hopkins, P.F., Richards, G.T., and Hernquist L., *Astrophys. J.* **654**, 731(2007).

Hubeny, I. *et al.*, *Astrophys. J.* **559**, 680 (2001).

Inayoshi, K., Haiman, Z. and Ostriker, J.P., *Mon. Not. Roy. Astron. Soc.* **459**, 3738 (2016).

Jiang, Y.-F., Stone, J. and Davis, S.W., arXiv:1709.02845 (2017).

Kaplan, S.A., *Dokl. Akad. Nauk. SSSR* **94**, 33 (1954).

Kaplan, S.A. and Pikel'ner, S.B., in *The Interstellar Medium* (Harvard University Press, Cambridge, 1970).

Katz, N., Weinberg, D.H., and Hernquist, L., *Astrophys. J. Supp.* **105**, 19 (1998).

Koide, S. *et al.*, *Science* **295**, 1688 (2002).

Kormendy, J. and Richstone, D., *Ann. Rev. Astron. Astrophys.* **33**, 581 (1995).

Kormendy, J. and Gebhardt, K., *AIP Conf. Proc.* **586**, 363 (2001).

Koushiappas, S.M., Bullock, J.S., and Dekel, A., *Mon. Not. Roy. Astron. Soc.* **354**, 292 (2004).

Levine, R. *et al.*, *Astrophys. J.* **678**, 154 (2008).

Magorrian, J. *et al.*, *Astron. J.* **115**, 2285 (1998).

Maraschi, L., Reina, C., and Treves, A., *Astron. Astrophys.* **35**, 389 (1974).

Marconi, A. and Hunt, L.K., *Astrophys. J.* **589**, L21 (2003).

Merrit, D. and Ferrarese, *Mon. Not. Roy. Astron. Soc.* **320**, L30 (2001).

Mihos, C. and Hernquist, L., *Astrophys. J.* **464**, 641 (1996).

Milosavljevic, M., Couch, S.M., and Bromm, V., *Astrophys. J.* **696**, L146 (2009).

Montesinos, M. and de Freitas Pacheco, J.A., *Astron. Astrophys.* **526**, A146 (2011).

Navarro, J.F. and White, S.D.M., *Mon. Not. Roy. Astron. Soc.* **265**, 271 (1993).

Nemmen, R.S. and Brotherton, M.S., *Mon. Not. Roy. Astron. Soc.* **408**, 1598 (2010).

Nobili, L., Turolla, R. and Zampieri, L., *Astrophys. J.* **383**, 250 (1991).

Pacucci, F., Volonteri, M. and Ferrara, A., *Mon. Not. Roy. Astron. Soc.* **452**, 1922 (2015).

Paumard, T. *et al.*, *Jour. Phys. Conf. Ser.* **54**, 199 (2006).

Peebles, P.J.E., in *Principles of Physical Cosmology* (Princeton University Press, Princenton, 1993).

Peirani, S. and de Freitas Pacheco, J.A., *Phys. Rev. D* **77**, 064023 (2008).

Piran, T., *Astrophys. J.* **221**, 652 (1978).

Regan, J.A. and Haehnelt, M.G., *Mon. Not. Roy. Astron. Soc.* **396**, 343 (2009).

Richstone, D. *et al.*, *Nature* **395**, 14 (1998).

Salpeter, E.E., *Astrophys. J.* **140**, 79 (1964).

Sazonov, S.Yu., Ostriker, J.P., Ciotti, L., and Sunyaev, R.A., *Mon. Not. Roy. Astron. Soc.* **358**, 168 (2005).

Schmidt, M., *Nature* **197**, 1040 (1963).

Shao, Y. *et al.*, *Astrophys. J.* **845**, 138 (2017).

Shakura, N.I. and Sunyaev, R.A., *Astron. Astrophys.* **24**, 337 (1973).

Shapiro, S.L., *Astrophys. J.* **613**, 1213 (2004).

Shvartsman, V.F., *Sov. Astron. (AJ)* **15**, 377 (1971).

Silk, J. and Rees, M.J., *Astron. Astrophys.* **331**, L1 (1998).

Small, T.A. and Blandford, R.D., *Mon. Not. Roy. Astron. Soc.* **259**, 725 (1992).

Soltan, A., *Mon. Not. Roy. Astron. Soc.* **200**, 115 (1982).

Springel, V., *Mon. Not. Roy. Astron. Soc.* **364**, 1105 (2005).

Sutherland, R.S. and Dopita, M.A., *Astrophys. J. Supp.* **88**, 253 (1993).

Trakhtenbrot, B., Volonteri M. and Natarajan P., *Astrophys. J.* **836**, L1 (2017).

Tremaine, S. *et al.*, *Astrophys. J.* **574**, 740 (2002).

Venemans, B.P. *et al.*, arxiv:1712.01886 (2017).

Walter, F. *et al.*, *Nature* **457**, 699 (2009).

Weiss, A. *et al.*, *ASP Conference Series* **375**, 25 (2007).

Wise, J.H., Turk, M.J., and Abel, T., *Astrophys. J.* **682**, 745 (2008).

Wu, X.-B. *et al.*, *Nature* **518**, 512 (2015).

Chapter 7

Astrophysical Aspects of General Relativistic Mass Twin Stars

David Blaschke[*,†,‡,1], David Edwin Alvarez-Castillo[*,2],
Alexander Ayriyan[§,¶,3], Hovik Grigorian[§,¶,‖,4],
Noshad Khosravi Largani[*,**,5], and Fridolin Weber[††,‡‡,6]

[*]*Bogoliubov Laboratory for Theoretical Physics*
Joint Institute for Nuclear Research
Joliot-Curie Street 6, 141980 Dubna, Russia
[†]*Institute of Theoretical Physics, University of Wroclaw*
Max Born Place 9, 50-204 Wroclaw, Poland
[‡]*National Research Nuclear University (MEPhI)*
Kashirskoe Shosse 31, 115409 Moscow, Russia
[§]*Laboratory for Information Technologies*
Joint Institute for Nuclear Research, Joliot-Curie Street 6
141980 Dubna, Russia
[¶]*Computational Physics and IT Division*
A.I. Alikhanyan National Science Laboratory
Alikhanyan Brothers Street 2, 0036 Yerevan, Armenia
[‖]*Department of Physics, Yerevan State University*
Alek Manukyan Street 1, 0025 Yerevan, Armenia
[**]*Department of Physics, Alzahra University*
Tehran, 1993893973, Iran
[††]*Department of Physics, San Diego State University*
5500 Campanile Drive, San Diego, CA 92182, USA
[‡‡]*Center for Astrophysics and Space Sciences*
University of California, San Diego, La Jolla, CA 92093, USA
[1]*david.blaschke@uwr.edu.pl*
[2]*sculkaputz@gmail.com*
[3]*ayriyan@jinr.ru*
[4]*hovikgrigorian@gmail.com*
[5]*noshad.khosravilargani@uwr.edu.pl*
[6]*fweber@sdsu.edu*

In this chapter we will introduce an effective equation of state (EoS) model based on polytropes that serves to study the so-called "mass twins" scenario, where two compact stars have approximately the same mass but (significant for observation) quite different radii. Stellar mass twin configurations are obtained if a strong first-order phase transition occurs in the interior of a compact star. In the mass–radius diagram of compact stars, this leads to a third branch of gravitationally stable stars with features that are very different from those of white dwarfs and neutron stars. We discuss rotating hybrid star sequences in the slow rotation approximation and in full General Relativity and draw conclusions for an upper limit on the maximum mass of non-rotating compact stars, which has recently been deduced from the observation of the merger event GW170817.

7.1. Introduction

Compact stars, the stellar remnants following the death of main sequence stars, have been the subject of investigation since the beginning of the last century. In particular, the determination of the internal composition of neutron stars is an open problem. Researching it involves many areas of physics, like nuclear, plasma, particle physics and relativistic astrophysics. Moreover, due to the enormous compactness (as expressed in the mass–radius ratio) of compact stars, these objects are extremely relativistic. Therefore, one can neither exclusively apply non-relativistic quantum mechanics nor classical Newtonian gravity to describe the observational properties of compact stars.

During the last decade important astronomical observations have shed light onto the nature of the dense, cold matter in the stellar interiors of compact stars. The detection of massive neutron stars, of about $2M_\odot$, has constrained the maximum density values in their cores and also revealed the stiff nature of the nuclear equation of state (EoS) at ultra-high densities. Strongly related to this issue, and one of the most interesting aspects of modern dense-matter physics, concerns the possible onset of quark deconfinement in the cores of compact stars.

Microscopic models that take into account the nuclear interactions either at the nucleon or quark level aim at providing a realistic hadronic or quark matter EoS, respectively. Neutron star matter must be thermodynamically consistent. Interestingly, due to the fast cooling of neutron stars after their birth in a supernova collapse, the thermal contributions to the EoS do not contribute substantially and can safely be neglected [Yakovlev *et al.* (2001)]. The thermodynamic system can therefore be described by three macroscopic variables: energy density ε, baryonic density n, and pressure P. A fourth quantity of great interest, the chemical potential, can then be obtained as $\mu = (P + \varepsilon)/n$.

In addition, the most basic conditions that the system must fulfill include global charge neutrality and β-equilibrium. The latter is derived from the reaction balance of beta decay and its inverse, the electron capture, due to the weak interactions. β-equilibrium fixes the relation between the chemical potentials of different species in the system.

With the above conditions satisfied, the neutron star equation of state becomes an expression of the form $P(\varepsilon)$, where $\varepsilon = \varepsilon(n)$ and $P = P(n)$ acquire parametric forms. Furthermore, in order to compute the internal properties of compact stars, it is necessary to obtain internal pressure profiles. This will result in mass–radius relations that characterize an EoS. Neutron stars are extremely relativistic objects which require to be treated within Einstein's general theory of relativity rather than simply Newtonian gravity, which may still be applicable to white dwarf stars. In this sense, our contribution addresses "strong gravity" in a unique fashion. To give an example, for a pulsar of mass $M = 2\,M_\odot$ with a typical radius of around $12\,\mathrm{km}$, the general relativistic correction factor amounts to $1/(1 - 2GM/R) = 2$ [Tolman (1939); Oppenheimer and Volkoff (1939)], which is a 100% correction relative to Newtonian gravity!

In the following sections we will introduce an effective EoS model based on polytropes (Alvarez-Castillo and Blaschke, 2017) that serves to study the "mass twins" scenario, where two compact stars have approximately the same mass but quite different radii, which has significant observational consequences [Glendenning and Kettner (2000)]. In the mass–radius diagram of compact stars, this will lead to a third branch of gravitationally stable stars with features that are very distinctive from those of white dwarfs and neutron stars.

The condition on the EoS that will lead to mass twins was first derived by Seidov (Seidov, 1971), see also (Schaeffer *et al.*, 1983;

Lindblom, 1998), namely that the central energy density ε_c, central pressure P_c, and the jump in energy associated with the phase transition $\Delta\varepsilon$ obey the relation

$$\frac{\Delta\varepsilon}{\varepsilon_c} \geq \frac{1}{2} + \frac{3}{2}\frac{P_c}{\varepsilon_c}. \tag{7.1}$$

When fulfilled, the corresponding compact star will suffer an instability of the same type as the maximum mass star of a stellar sequence. Most interestingly, stars with central densities higher than the density of the maximum-mass star become stable again if their gravitational masses obey $\partial M/\partial\varepsilon_c(0) > 0$, thereby populating a third branch with stable mass twins.

7.2.   Self-Consistent Set of Field Equations for Stationary Rotating and Tidally Deformed Stars

The geometrical description of the spacetime structure curved due to the mass–energy of a compact star is given by the general metric form defining the interval between the infinitesimally close events,

$$ds^2 = g_{\mu\nu}(x)dx^\mu dx^\nu. \tag{7.2}$$

The curvature of spacetime satisfies the Einstein field equations $G_\nu^\mu = 8\pi G T_\nu^\mu$, where $G_\nu^\mu = R_\nu^\mu - \delta_\nu^\mu R/2$. Here R_ν^μ is the Ricci curvature tensor and R the scalar curvature. On the right hand side of the Einstein equation we have the energy–momentum tensor T_ν^μ of the stellar matter and G is the gravitational constant ($\hbar = c = 1$).

The metric tensor $g_{\mu\nu}(x)$ has the same symmetry as the matter distribution. Therefore, if one assumes that the star is static and not deformed the metric tensor is diagonal and depends only on the distance from the center of the star. In the case of stationary rotating stars the symmetry of the matter will be axially symmetric. In this case due to the rotational motion of the star the non-diagonal element $g_{t\phi}$ (t is the time coordinate, ϕ the azimuthal angle of the spherical coordinate system) will not be zero in the inertial frames connected with the star. The existence of such a term leads to the Lense–Thirring effect of frame dragging for the motion of bodies in the gravitational field of rotating compact relativistic stellar objects.

However, in the case of small but static deformations the metric will be non-spherical but diagonal.

7.2.1. *Einstein equations for axial symmetry*

The general form of the metric for an axially symmetric spacetime manifold in the inertial frame where the star center is at rest is

$$ds^2 = e^{\nu(r,\theta)} dt^2 - e^{\lambda(r,\theta)} dr^2 - r^2 e^{\mu(r,\theta)} [d\theta^2 + \sin^2 \theta (d\phi + \omega(r,\theta) dt)^2],$$

$$(7.3)$$

where a spherically symmetric coordinate system has been used in order to obtain the Schwarzschild solution as a limiting case. This line element is time, translation and axial-rotational invariant; all metric functions are dependent on the coordinate distance from the coordinate center r and altitude angle θ between the radius vector and the axis of symmetry.

The energy–momentum tensor of stellar matter can be approximated by the expression of the energy–momentum tensor of an ideal fluid

$$T_\mu^\nu = (\varepsilon + P) u_\mu u^\nu - P \delta_\mu^\nu, \tag{7.4}$$

where u^μ is the 4-velocity of matter, P the pressure and ε the energy density.

Once the energy–momentum tensor (7.4) is fixed by the choice of the equation of state for stellar matter, the unknown metric functions ν, λ, μ, $\bar{\omega}$ can be determined by the set of Einstein field equations for which we use the following four combinations.

There are three Einstein equations for the determination of the diagonal elements of the metric tensor,

$$G_r^r - G_t^t = 8\pi G(T_r^r - T_t^t), \tag{7.5}$$

$$G_\theta^\theta + G_\phi^\phi = 8\pi G(T_\theta^\theta + T_\phi^\phi), \tag{7.6}$$

$$G_\theta^r = 0, \tag{7.7}$$

and one for the determination of the non-diagonal element

$$G_\phi^t = 8\pi G T_\phi^t. \tag{7.8}$$

We use also one equation for the hydrodynamic equilibrium (Euler equation)

$$H(r,\theta) \equiv \int \frac{dP'}{P' + \varepsilon'} = \frac{1}{2} \ln[u^t(r,\theta)] + \text{const.}, \qquad (7.9)$$

where the gravitational enthalpy H thus introduced is a function of the energy and/or pressure distribution.

7.2.2. *Full solution for uniform rotational bodies*

In this section we describe the method of solution employed by the RNS code written by Stergioulas and Friedman (1995), based on the method developed by Komatsu *et al.* (1989) that also includes modifications by Cook *et al.* (1994). In addition, the inclusion of quadrupole moments is due to Morsink based on the method by Laarakkers and Poisson (1999).

In order to study the full solutions for rotating compact stars the following metric is considered [Cook *et al.* (1994)]:

$$ds^2 = -e^{\gamma(r,\theta)+\rho(r,\theta)}dt^2 + e^{2\alpha(r,\theta)}(dr^2 + r^2 d\theta^2)$$
$$+ e^{\gamma(r,\theta)-\rho(r,\theta)}r^2 \sin^2\theta(d\phi - \omega(r,\theta)dt)^2, \qquad (7.10)$$

which just like Eq. (7.3) properly describes a stationary, axisymmetric spacetime. In addition, the matter source is chosen to be a perfect fluid described by the stress–energy tensor $T^{\mu\nu} = (\epsilon+P)u^\mu u^\nu + Pg^{\mu\nu}$, where u^μ is the four-velocity of matter. Three of the solutions to the gravitational field equations are found by a Green function approach therefore leading to the determination of the metric potentials ρ, γ and ω in term of integrals, whereas the α potential is found by solving a linear differential equation. Therefore, we find the corresponding numerical solutions to our compact star models by employing the RNS code. In the formulation of the problem, all the physical variables are written in dimensionless form by means of a fundamental length scale $\sqrt{\kappa}$, where $\kappa \equiv \frac{c^2}{G\epsilon_0}$ with $\epsilon_0 \equiv 10^{15}\,\mathrm{g\,cm^{-3}}$. The dimensionless variables are given in Table 7.1. The output parameters of the code are given in Table 7.2.

Table 7.1 Dimensionless physical variables used in the RNS code formulation. $\kappa \equiv c^2/(G\epsilon_0)$, $\epsilon_0 = 10^{15}\,\mathrm{g\,cm^{-3}}$.

$\bar{r}$	$\kappa^{-1/2} r$
$\bar{t}$	$\kappa^{-1/2} ct$
$\bar{\omega}$	$\kappa^{1/2} \dfrac{1}{c} \omega$
$\bar{\Omega}$	$\kappa^{1/2} \dfrac{1}{c} \Omega$
$\bar{\rho}_0$	$\kappa \dfrac{G}{c^2} \rho_0$
$\bar{\epsilon}$	$\kappa \dfrac{G}{c^2} \epsilon$
$\bar{P}$	$\kappa \dfrac{G}{c^4} P$
$\bar{J}$	$\kappa^{-1} \dfrac{G}{c^3} J$
$\bar{M}$	$\kappa^{-1/2} \dfrac{G}{c^2} M$

Table 7.2 Output parameters of the RNS code.

Gravitational mass–energy	$M/M_\odot$
Rest mass	$M_0/M_\odot$
Circumferencial radius [km]	R_e
Eccentricity	e
Central energy density $[10^{15}\,\mathrm{g\,cm^{-3}}]$	ϵ_c
Angular velocity measured at infinity $[10^3\,\mathrm{s^{-1}}]$	Ω
Total angular momentum	$cJ/GM_\odot^2$
Rotational kinetic energy over gravitational energy	T/W
Measure of frame dragging	ω_c/Ω_c
Polar redshift	Z_p
Equatorial redshift in backward direction	Z_b
Equatorial redshift in forward direction	Z_f
Circumferential height of corotating marginally stable orbit [km]	h_+
Circumferential height of counterrotating marginally stable orbit [km]	h_-

The global parameters of the star are computed by means of the following expressions:

$$M = \frac{4\pi\kappa^{1/2}c^2\bar{r}_e^3}{G} \int_0^1 \frac{s^2 ds}{(1-s)^4} \int_0^1 d\mu \; e^{2\alpha+\gamma}$$

$$\times \left\{ \frac{\bar{\epsilon}+\bar{P}}{1-v^2}\left[1+v^2+\frac{2sv}{1-s}(1-\mu)^{1/2}\hat{\omega}e^{-\rho}\right]+2\bar{P}\right\}, \quad (7.11)$$

$$M_0 = \frac{4\pi\kappa^{1/2}c^2\bar{r}_e^3}{G} \int_0^1 \frac{s^2 ds}{(1-s)^4} \int_0^1 d\mu \; e^{2\alpha+(\gamma-\rho)/2} \frac{\bar{\rho}_0}{(1-v^2)^{1/2}}, \quad (7.12)$$

$$J = \frac{4\pi\kappa c^3\bar{r}_e^4}{G} \int_0^1 \frac{s^3 ds}{(1-s)^5} \int_0^1 d\mu(1-\mu^2)^{1/2}\; e^{2\alpha+\gamma-\rho}(\bar{\epsilon}+\bar{P})\frac{v}{1-v^2},$$
$$(7.13)$$

$$T = \frac{2\pi\kappa^{1/2}c^2\bar{r}_e^3}{G} \int_0^1 \frac{s^3 ds}{(1-s)^5}$$

$$\times \int_0^1 d\mu(1-\mu^2)^{1/2}\; e^{2\alpha+\gamma-\rho}(\bar{\epsilon}+\bar{P})\frac{v\hat{\Omega}}{1-v^2}, \quad (7.14)$$

where $\hat{\omega} \equiv \bar{r}_e\bar{\omega}$ and $\hat{\Omega} \equiv \bar{r}_e\bar{\Omega}$ with $\bar{r}_e$ as the coordinate radius of the equator. Moreover, all the resulting quantities can be written in terms of the auxiliary variables μ and s, defined as $\mu \equiv \theta$ and $\bar{r} \equiv \bar{r}_e\left(\frac{s}{1-s}\right)$, respectively. Consequently, the four metric functions acquire a dependence on the above variables: $\rho(s,\mu)$, $\gamma(s,\mu)$, $\omega(s,\mu)$, $\alpha(s,\mu)$. The remaining quantities are:

$$R_e = \kappa^{1/2}\bar{r}_e e^{(\gamma_e-\rho_e)/2}, \quad (7.15)$$

$$Z_p = e^{-(\gamma_p+\rho_p)/2} - 1, \quad (7.16)$$

$$Z_f = \left(\frac{1-v_e}{1+v_e}\right)^{1/2} \frac{e^{-(\gamma_e+\rho_e)/2}}{1+\hat{\omega}_e e^{-\rho_e}} - 1, \quad (7.17)$$

$$Z_b = \left(\frac{1+v_e}{1-v_e}\right)^{1/2} \frac{e^{-(\gamma_e+\rho_e)/2}}{1-\hat{\omega}_e e^{-\rho_e}} - 1, \quad (7.18)$$

where the subscripts e and p denote evaluation at the equator and at the pole, respectively.

For the solutions of maximally rotating compact stars in numerical general relativity, the version of RNS code employed is available for download from the website http://www.gravity.phys.uwm.edu/rns/.

7.2.3. *Perturbation approach to the solution*

The problem of the rotation can be solved iteratively by using a perturbation expansion of the metric tensor and the physical quantities in a Taylor series with respect to a small positive parameter β. As such a parameter for the perturbation expansion we use a dimensionless quantity. One possible physically motivated way is to take the ratio of the rotational or deformation energy to the gravitational one. The gravitational energy could be estimated for a homogeneous Newtonian star as $\beta = E_{\text{def}}/E_{\text{grav}}$. In case of rotating stars the deformability is connected with the induced centrifugal force and the expansion parameter is $\beta = E_{\text{rot}}/E_{\text{grav}} = (\Omega/\Omega_0)^2$, where $\Omega_0^2 = 4\pi G\rho(0)$ with the mass density $\rho(0)$ at the center of the star. The choice of this parameter could be also motivated by the conditions of the problem. For example since the stationary rotating stars cannot have too high value of the angular velocity, because of mass shedding on Keplerian angular velocity $\Omega_K = \sqrt{GM/R_e^3}$ for the star with total mass M and R_e equatorial radius, the expansion gives sufficiently correct solutions already at $O(\Omega^2)$. So the expansion parameter is naturally limited to values $\Omega/\Omega_0 \ll 1$ by this condition of mechanical stability of the rigid rotation, because always $\Omega < \Omega_K = \Omega_0/\sqrt{3}$. This condition is fulfilled not only for homogeneous Newtonian spherical stars but also for the relativistic configurations even with a possible hadron–quark (deconfinement) transition, which we are going to discuss later in Sec. 7.5.1. The perturbation approach to slowly rotating stars has been developed first by Hartle (1967); Hartle and Thorne (1968), and independently by Sedrakyan and Chubaryan (1968a,b) and is described in detail in [Weber, F. (1999); Glendenning, N. K. (2000); Chubarian *et al.* (2000)]. Our notation and derivation in this section will follow Chubarian *et al.* (2000) while numerical solutions for the slow rotation (Ω^2-) approximation are obtained with a code based on the improved Hartle scheme [Weber and Glendenning (1992)].

The expansion of the metric tensor in a perturbation series with respect to the slow rotation parameter β can be expressed as

$$g_{\mu\nu}(r,\theta) = \sum_{j=0}^{\infty} \left(\sqrt{\beta}\right)^j g_{\mu\nu}^{(j)}(r,\theta) \,. \tag{7.19}$$

According to the metric form for the axial symmetry in the linear approximation via β parameter we introduce the notations describing explicitly the non-perturbed (spherically symmetric case noted with upper index (0)) and perturbed terms (corresponding to $j = 1, 2$) in the metric,

$$
\begin{aligned}
e^{-\lambda(r,\theta)} &= e^{-\lambda^{(0)}(r)}[1 + \beta f(r,\theta)] + O(\beta^2), \\
e^{\nu(r,\theta)} &= e^{\nu^{(0)}(r)}[1 + \beta\Phi(r,\theta)] + O(\beta^2), \\
e^{\mu(r,\theta)} &= r^2[1 + \beta U(r,\theta)] + O(\beta^2),
\end{aligned} \tag{7.20}
$$

and for the frame dragging frequency ω the odd orders

$$\omega(r,\theta) = \sqrt{\beta}q(r,\theta) + O((\sqrt{\beta})^3). \tag{7.21}$$

In the same way one performs a velocity expansion of the energy–momentum tensor, the pressure and energy density distributions, and of the kinetic energy,

$$P(r,\theta) = P^{(0)}(r) + \beta P^{(2)}(r,\theta) + O(\beta^2), \tag{7.22}$$

$$\varepsilon(r,\theta) = \varepsilon^{(0)}(r) + \beta\varepsilon^{(2)}(r,\theta) + O(\beta^2), \tag{7.23}$$

where $P^{(0)}$ and $\varepsilon^{(0)}$ denote the zeroth-order coefficients which correspond to non-deformed spherically symmetric stars.

Because of rotational symmetry the diagonal elements $g_{\mu\mu}^{(j)}(r,\theta)$ (no summation over μ) of the metric coefficients can be written as (j and l are even values only)

$$g_{\mu\mu}^{(j)}(r,\theta) = \sum_{l=0}^{j}(g_{\mu\mu})_l(r)P_l(\cos\theta) \,. \tag{7.24}$$

The same is true also for the non-diagonal elements, but the angular dependence is different (j and l values are now odd only). The case for the non-diagonal term will be investigated in Sec. 7.3.1, where the moment of inertia will be discussed.

In the case of a deformed distribution of the matter, the external metric has the form of the Kerr metric. However this solution is for black holes and does not correspond to a realistic stellar model. Therefore the external as well as the internal solutions of the field and matter distributions can only be obtained either via a perturbation approximation or via a completely numerical treatment.

7.2.4. *Static spherically symmetric star models*

As a first step in the perturbation approach of a slowly rotating star one needs to find the internal gravitational field, the mass and matter distributions, the total gravitational mass, the radius and all other characteristic properties (including the metric functions) of spherically symmetric stars. The solution for the metric functions in empty space (i.e., the external solutions) are given by the Schwarzschild solution,

$$\lambda^{(0)}(r) = -\ln[1 - 2GM/r],$$
$$\nu^{(0)}(r) = -\lambda^{(0)}(r), \tag{7.25}$$

where M is a constant of integration, which asymptotically is the Newtonian gravitational mass of the object.

These nonlinear equations, however, could be written in an elegant form suggested by Tolman, Oppenheimer and Volkoff, which are known as the TOV equations (Tolman, 1939; Oppenheimer and Volkoff, 1939), and solved in a way such that the internal solution matches the analytic external Schwarzschild solution. The TOV equation is given by (for a derivation, see, e.g., the textbook by Misner *et al.* (1973))

$$\frac{dP^{(0)}(r)}{dr} = -G[P^{(0)}(r) + \varepsilon^{(0)}(r)]\frac{m(r) + 4\pi P^{(0)}(r)r^3}{r[r - 2Gm(r)]}, \tag{7.26}$$

where $P^{(0)}$ and $\varepsilon^{(0)}$ denote the equation of state (EoS) describing the stellar matter. The quantity $m(r)$, defined as

$$m(r) = 4\pi \int_0^r \varepsilon^{(0)}(r')r'^2 dr', \tag{7.27}$$

stands for the amount of gravitational mass contained inside a sphere of radius r, with r denoting the distance from the center of the star. The star's total gravitational mass, M, is then given by

$$M = m(R) = 4\pi \int_0^R \varepsilon^{(0)}(r')r'^2 dr', \qquad (7.28)$$

where R denotes the radius of the star defined by $P(r = R) = 0$. Physically, the TOV equation describes the balance of gravitational and internal pressure forces at each radial distance inside the star. Both forces exactly cancel each other inside a static stellar configuration, as described by the TOV equation.

The TOV equation is solved numerically, for a given model for the EoS, by choosing a value for the star's central density and then integrating Eq. (7.26) out to a radial location where the pressure becomes zero. So for any fixed choice of the EoS, the stars form a one-parameter sequence (parameter ε_c^0). An entire family of compact stars is obtained by solving the TOV equation for a range of central densities which result in the mass–radius relationship of compact stars. It is characterized by the existence of a maximum mass star (several maximum mass stars if permitted by the model chosen for the EoS). The stars are stable against gravitational collapse if they are on the stellar branch for which $\partial M/\partial \varepsilon_c^{(0)} > 0$. Stars on the stellar branch where $\partial M/\partial \varepsilon_c^{(0)} < 0$ are unstable against radial oscillations and will therefore not exist stably in the universe.

Each stellar model has unique solutions for $m(r)$, $P^{(0)}(r)$ and $\varepsilon^{(0)}(r)$ in terms of which the internal gravitational field (the metric coefficients) is defined as

$$\lambda^{(0)}(r) = -\ln[1 - 2Gm(r)/r], \qquad (7.29)$$

$$\nu^{(0)}(r) = -\lambda^{(0)}(R_0) - 2G \int_r^{R_0} \frac{m(r') + 4\pi P^{(0)}(r')r'^3}{r'[r' - 2Gm(r')]} dr'. \qquad (7.30)$$

The internal field solutions are smoothly connected to the external field solutions at the stellar surface, $r = R$. Once the internal pressure profiles are derived from the solution of the TOV equations, it is possible to compute other astrophysical quantities like baryonic mass, and also make the next step in the perturbation approach to define the moment of inertia and tidal deformabilities, which are of very great observational interest.

Of particular interest for astrophysical scenarios (stellar evolution) is the expression for the total baryon mass

$$\frac{dN_B(r)}{dr} = 4\pi r^2 \left(1 - \frac{2Gm(r)}{r}\right)^{-1/2} n(r), \qquad (7.31)$$

where $n(r)$ is the baryon number density and $N_B(R)$ is the total baryon number of the star. This number is a characteristic conserved quantity and is very important in discussions of evolutionary scenarios of compact stars, see, e.g., Bejger *et al.* (2017); Ayvazyan *et al.* (2013); Chubarian *et al.* (2000); Poghosyan *et al.* (2001). The functions of the spherically symmetric solution in Eqs. (7.20) and (7.22) can be found from Eqs. (7.5) and (7.9) in zeroth order of the Ω-expansion.

7.3. Tidal Deformability of Compact Stars

The tidal deformability (TD) is a measure of the shape deformation property of the astrophysical object under the gravitational influence of another nearby object. To determine it in the first order we need to consider a modification of the spacetime metric when the distribution of matter of the star becomes elliptic. According to the symmetries of the metric coefficients introduced in Eq. (7.3) we have even orders $j = 0, 2, \ldots$ for the diagonal elements.[1] For the first correction corresponding to small deformations (small values of β) one can consider terms linear in β or equivalently the $j = 2$ perturbation approximation to the spherically symmetric star. We introduce some new notation and work under the assumption that $f_2(r) = -\Phi_2(r) = A(r)$ like in the expansion of the external solution, since $\nu^{(0)}(r) = -\lambda^{(0)}(r)$.

The non-diagonal term could be taken to be zero, because we consider only the static case $q(r, \theta) = 0$ (the parameter defining the static deformation $\sqrt{\beta}$ does not change the sign under time reversal

---

[1]Notation corresponds to the work of Chubarian *et al.* (2000).

$t \to -t)$ and $U_2(r) = K(r)$, so that we have

$$ds^2 = e^{\lambda^{(0)}(r)}[1 + \beta A(r)P_2(\theta)]dt^2$$
$$- e^{\nu^{(0)}(r)}[1 - \beta A(r)P_2(\theta)]dr^2$$
$$- r^2[1 - \beta K(r)P_2(\theta)](d\theta^2 + \sin^2\theta d\varphi^2). \tag{7.32}$$

In this approximation, without loss of generality, one can set the values of $f_0 = \Phi_0 = U_0 = 0$, because we neglect the contribution of the deformation energy to the gravitational mass. The equations show that $K'(r) = A'(r) + \nu^{(0)'}(r)A(r)$, the prime symbol denoting the derivative of those quantities with respect to r. The functions $A(r)$ and $B(r) = dA/dr$ obey the differential equations

$$\frac{dA(r)}{dr} = B(r); \tag{7.33}$$

$$\frac{dB(r)}{dr} = 2\left(1 - 2G\frac{m(r)}{r}\right)^{-1}$$
$$\times A(r)\left\{-2\pi\left[5\varepsilon^{(0)}(r) + 9P^{(0)}(r) + \frac{1}{c_s^2}(\varepsilon^{(0)}(r) + P^{(0)}(r))\right]\right.$$
$$+ \frac{3}{r^2} + 2\left(1 - 2G\frac{m(r)}{r}\right)^{-1}\left[G\left(\frac{m(r)}{r^2} + 4\pi r P^{(0)}(r)\right)\right]^2\right\}$$
$$+ \frac{2B(r)}{r}\left(1 - 2G\frac{m(r)}{r}\right)^{-1}$$
$$\times\left\{-1 + G\left[\frac{m(r)}{r} + 2\pi r^2\left(\varepsilon^{(0)}(r) - P^{(0)}(r)\right)\right]\right\}. \tag{7.34}$$

Here, $c_s^2 = dP/d\varepsilon$ is the square of the speed of sound, which is equivalent to the knowledge of the equation of state. The pressure profile provided by solving the TOV equations will complement the above equations.

The system is to be integrated with the asymptotic behavior of metric functions $A(r) = a_0 r^2$ and $B(r) = 2a_0 r$ as $r \to 0$. The a_0 is a constant that quantifies the deformation of the star which can be taken as arbitrary. This constant corresponds to the choice of β

as an external parameter. Since it cancels in the expression for the Love number and in all other quantities in consideration, its value is not important. Using the solution on the surface at $r = R$ and the following combination,

$$y = \frac{R\,B(R)}{A(R)}, \tag{7.35}$$

it is possible to compute the $l = 2$ Love number (Hinderer, 2008; Damour and Nagar, 2009; Binnington and Poisson, 2009; Yagi and Yunes, 2013; Hinderer *et al.*, 2010):

$$
\begin{aligned}
k_2 ={}& \frac{8C^5}{5}(1 - 2C)^2[2 + 2C(y - 1) - y] \\
&\times \{2C[6 - 3y + 3C(5y - 8)] \\
&+ 4C^3[13 - 11y + C(3y - 2) + 2C^2(1 + y)] \\
&+ 3(1 - 2C)^2[2 - y + 2C(y - 1)]\ln(1 - 2C)\}^{-1},
\end{aligned}
$$

where M/R in the expression $C = GM/R$ is the compactness of the star ($2C$ is the ratio of gravitational radius to spherical radius).

The dimensionless tidal deformability parameter is defined as $\Lambda = \lambda/M^5$, a quantity defined for small tidal deformabilities. Here λ is the TD of the star with a gravitational mass M, just as defined above. In addition, the Love number is related to TD and defined as

$$k_2 = \frac{3}{2}\lambda R^{-5}. \tag{7.36}$$

In the investigations and observations of the process of neutron star merging, the TD Λ is a key parameter characterizing the stiffness of equation of state of the stellar matter.

7.3.1. *Moment of inertia*

The moment of inertia is one of the main characteristics of the mechanical properties of the rotating body, therefore one needs to define it also for the relativistic objects (such as neutron stars) obeying the gravitational field contribution to the rotational motion.

The baryonic mass is an important quantity often associated with explosive events, where it can be conserved while the gravitational mass of the star suffers modifications (Alvarez-Castillo *et al.*, 2015; Bejger *et al.*, 2017). The moment of inertia is related to the *glitch* phenomenon, which is a sudden spin-up in the general spin-down evolution of rotation frequencies, observed for some pulsars, see (Haskell and Melatos, 2015) and references therein. Moreover, it is expected to be measured in pulsar binaries, providing a strong constraint on the compact star EoS (Lattimer and Prakash, 2007).

In a very simplified way one can estimate the impact of relativistic effects on the moment of inertia from (Ravenhall and Pethick, 1994)

$$I \simeq \frac{C_J}{1 + 2GJ/R^3},\tag{7.37}$$

where J denotes the total, conserved angular moment. The quantity C_J is given by

$$C_J = \frac{8\pi}{3} \int_0^R r^4 (\varepsilon^{(0)}(r) + P^{(0)}(r)) \frac{1}{1 - 2Gm(r)/r} dr,\tag{7.38}$$

which can be readily computed since only the knowledge of spherically symmetric quantities is required, which are easy to compute.

However, because of the deformation of the star due to rotation and the impact of the gravitational field on the rotational inertia of the star, the moment of inertia becomes a function of the rotational state, i.e., a function of the angular velocity or spin frequency of the neutron star. To take all these effects into account in the defining expression for the moment of inertia we will follow the steps of our perturbation approach.

By definition, the angular momentum of the star in the case of stationary rotation is a conserved quantity and can be expressed in invariant form

$$J = \int T^t_\phi \sqrt{-g} dV,\tag{7.39}$$

where $\sqrt{-g} dV$ is the invariant volume and $g = \det \|g_{\mu\nu}\|$. For the case of slow rotation where the shape deformation of the rotating star can be neglected and using the definition of the moment of inertia $I_0(r) = J_0(r)/\Omega$ accumulated in the sphere with radius r, we obtain from Eq. (7.39)

$$\frac{dI_0(r)}{dr} = \frac{8\pi}{3} r^4 [\varepsilon^{(0)}(r) + P^{(0)}(r)] e^{(-\nu^{(0)}(r) + \lambda^{(0)}(r))/2} \frac{\bar{\omega}(r)}{\Omega}.$$

Here $\bar{\omega}$ is the difference between the frame dragging frequency $-\omega$ and the angular velocity Ω. In General Relativity, due to the Lense–Thirring law, rotational effects are described by

$$\bar{\omega} \equiv \Omega + \omega(r, \theta). \tag{7.40}$$

This expression is approximated from the exact expression of the energy–momentum tensor coefficient T^t_ϕ and the metric tensor in the axial symmetric case.

We keep only two non-vanishing components of the 4-velocity

$$u^\phi = \Omega \, u^t,$$
$$u^t = 1/\sqrt{e^\nu - r^2 e^\mu \bar{\omega}^2 \sin^2 \theta}, \tag{7.41}$$

because we assume that the star due to high viscosity (ignoring the super-fluid component of the matter) rotates as a solid body with an angular velocity Ω that is independent of the spatial coordinates. The time scales for changes in the angular velocity which we will consider in our applications are well separated from the relaxation times at which hydrodynamical equilibrium is established, so that the assumption of a rigid rotator model is justified.

Now besides central energy density $\varepsilon(0)$ of the star configuration the angular velocity of the rotation Ω is an additional parameter of the theory.

As a next step, going beyond the spherically symmetric case that corresponds to the first-order approximation, we solve Eq. (7.8), where the unknown function is $q(r, \theta)$ which is defined by Eq. (7.21) and scaled such that it is independent of the angular velocity. Using the static solutions Eqs. (7.26) and (7.29), and the representation of

$q(r, \theta)$ by the series of the Legendre polynomials,

$$q(r, \theta) = \sum_{m=0}^{\infty} q_m(r) \frac{dP_{m+1}(\cos \theta)}{d \cos \theta}. \tag{7.42}$$

One can see that this series is truncated and only the coefficient $q_0(r)$ is nontrivial, i.e., $q_m(r) = 0$ for $m > 0$, see (Hartle, 1967; Chubarian *et al.*, 2000). Therefore one can write down the equations for $\bar{\omega}(r) = \Omega(1 + q_0(r)/\Omega_0)$, which is more suitable for the solution of the resulting equation in first order

$$\frac{1}{r^4} \frac{d}{dr} \left[r^4 j(r) \frac{d\bar{\omega}(r)}{dr} \right] + \frac{4}{r} \frac{dj(r)}{dr} \bar{\omega}(r) = 0, \tag{7.43}$$

which corresponds to Hartle (1967), where it was obtained using a different representation of the metric. In this equation we use the notation $j(r) \equiv e^{-(\nu^{(0)}(r) + \lambda^{(0)}(r))/2}$, where $j(r) = 1$ for $r > R_0$, i.e., outside of stellar configuration.

Using this equation one can reduce the second-order differential equation (7.43) to the first-order one

$$\frac{d\bar{\omega}(r)}{dr} = \frac{6G J_0(r)}{r^4 j(r)}. \tag{7.44}$$

and solve Eq. (7.43) as a coupled set of first-order differential equations, one for the moment of inertia (7.39) and the other (7.44) for the frame dragging frequency $\bar{\omega}(r)$.

This system of equations is valid inside and outside the matter distribution. At the center of the configuration $I_0(0) = 0$ and $\bar{\omega}(0) = \bar{\omega}_0$. The finite value $\bar{\omega}_0$ has to be defined such that the dragging frequency $\bar{\omega}(r)$ smoothly joins the outer solution

$$\bar{\omega}(r) = \Omega \left(1 - \frac{2G I_0}{r^3} \right) \tag{7.45}$$

at $r = R_0$, and approaches Ω in the limit $r \to \infty$. In the external solution (7.45) the constant $I_0 = I_0(R_0)$ is the total moment of inertia of the slowly rotating star and $J_0 = I_0\Omega$ is the corresponding angular momentum. In this order of approximation, I_0 is a function of the central energy density or the total baryon number only. This solution remains connected to the spherically distributed matter and therefore does not differ too much from our previous expression, which uses this solution to incorporate the relativistic corrections.

However, to find the explicit dependence of the moment of inertia on the angular velocity, one needs to take the second step and account for the deformation of the stellar configuration, which, in the framework of our scheme, is a second-order correction.

7.3.2. *Rotational deformation and moment of inertia*

To calculate these contributions and the internal structure of the rotating star which is deformed due to centrifugal force, one needs to return to our perturbation description and expand the diagonal elements of the metric and the energy–momentum tensor with the parameter $\beta = (\Omega/\Omega_0)^2$.

For a more detailed description of the solutions of the field equations in the $\sim O(\Omega^2)$ approximation we refer to the works of Hartle (1967) as well as Chubarian *et al.* (2000). Since these equations have a complicated form, here we will discuss only the qualitative meaning of the physical quantities concerning the star's deformation and its moment of inertia.

In Ω^2-approximation the shape of the star is an ellipsoid, and each of the equal-pressure (isobar) surfaces in the star is an ellipsoid as well. All diagonal elements of the metric and energy–momentum tensor could be represented as a series expansion in Legendre polynomials, as we have already discussed in the previous section. As noted, the only non-vanishing solutions obeying the continuity conditions on the surface with the external solution of fields are those with $l = 0, 2$.

The deformation of the isobaric surfaces can be parameterized by the deformation shifts $R(r, \theta) - r = \Delta(r, \theta)$ from the spherical shape. It describes the deviation from the spherical distribution as a function of radius r for a fixed polar angle θ and is completely determined by

$$R(r, \theta) = r + \left(\frac{\Omega}{\Omega_0}\right)^2 [\Delta_0(r) + \Delta_2(r)P_2(\cos\theta)], \qquad (7.46)$$

since the expansion coefficients of the deformation $\Delta_l(r)$ are connected to the pressure corrections

$$\Delta_l(r) = -\frac{p^{(l)}(r)}{dP^{(0)}(r)/dr}. \tag{7.47}$$

$l \in \{0, 2\}$ is the polynomial index in the angular expansion in Legendre polynomials, analogous to Eq. (7.24). The function $R(R_0, \theta)$ is the radius where $p(R(R_0, \theta)) = 0$. R_0 is the spherical radius, which is no longer the actual radius. The surface of the star is given by $R(R_0, \theta)$ measured from the center of the configuration at a polar angle θ. In particular, we define the equatorial radius as $R_e = R(R_0, \theta = \pi/2)$, the polar radius as $R_p = R(R_0, \theta = 0)$, and the eccentricity as $\epsilon = \sqrt{1 - (R_p/R_e)^2}$, all three quantities characterizing the deformed shape of the star.

Using the same approach we write the correction to the moment of inertia as $\Delta I(r) = I(r) - I_0(r)$ and represent it as a sum of several different contributions,

$$\Delta I = \Delta I_{\text{Redist.}} + \Delta I_{\text{Shape}} + \Delta I_{\text{Field}} + \Delta I_{\text{Rotation}}. \tag{7.48}$$

Since these contributions are obtained from the exact expression of angular momentum in the integral form, the first three contributions can also be expressed by integrals of the form

$$\Delta I_\alpha = \int_0^{I_0(R_0)} dI_0(r)[W_0^{(\alpha)}(r) - W_2^{(\alpha)}(r)/5], \tag{7.49}$$

where integration is taken from the angular averaged modifications of the matter distribution, the shape of the configuration and the gravitational fields,

$$W_l^{(\text{Field})}(r) = \left(\frac{\Omega}{\Omega_0}\right)^2 \{2U_l(r) - [f_l(r) + \Phi_l(r)]/2\}, \tag{7.50}$$

$$W_l^{(\text{Shape})}(r) = \left(\frac{\Omega}{\Omega_0}\right)^2 \frac{d\,\Delta_l(r)}{dr}, \tag{7.51}$$

$$W_l^{(\text{Redist.})}(r) = \left(\frac{\Omega}{\Omega_0}\right)^2 \frac{p_l(r) + \varepsilon_l(r)}{p^{(0)}(r) + \varepsilon^{(0)}(r)}, \tag{7.52}$$

respectively. All quantities appearing here are determined from Eq. (7.5) in second-order approximation. The contribution of the change of the rotational energy to the moment of inertia is given by

$$\Delta I_{\text{Rotation}} = \frac{4}{5} \int_0^{I_0(R_0)} dI_0(r)[r^2 \bar{\omega}^2(r)e^{-\nu_0(r)}], \qquad (7.53)$$

and includes the frame dragging contribution.

In the next sections of this chapter we will discuss results for the moment of inertia along with stability conditions. We note that a consistent discussion of the stability of rotating stars requires one to take into account the contribution of the rotational energy to the mass–energy, as well as the corresponding corrections to the moment of inertia.

7.4. Models for the EoS with a Strong Phase Transition

In this contribution, we focus on EoS models which describe a strong phase transition in the sense that upon solving the TOV equations, compact star sequences obtained exhibit a third family branch in the mass–radius or mass–central (energy) density diagram which is separated from the second family of neutron stars by a sequence of unstable configurations. The possibility of the very existence of a third family of compact stars as a consequence of a strong phase transition in dense nuclear matter, together with a sufficient stiffening of the high-density matter that can be expressed by a strong increase in the speed of sound (but not violating the causality bound) has been discussed by Gerlach as early as 1968 [Gerlach (1968)]. Let us note here that the existence of such a third family of compact stars is an effect of strong, general relativistic gravity! Namely, that the compactification which accompanies the strong phase transition of the star leads to a reduction of the gravitational mass of the hybrid star configuration from which it only recovers (and thus escapes gravitational collapse) when after the transition the hybrid star core consists of sufficiently stiff high-density matter.

Such EoS lead to mass–radius relationships for hybrid stars that have been classified (D)isconnected or (B)oth in [Alford *et al.* (2013)].

The "D" topology consists of a hadronic and a hybrid star branch, both of which being gravitationally disconnected from each other. In contrast to this, the "B" topology consists of a branch of stable hadronic stars followed by stable hybrid stars, which are gravitationally disconnected from a second branch of stable hybrid stars. For the introduction of this classification scheme, the constant-speed-of-sound (CSS) EoS was used in [Alford *et al.* (2013)] for describing the high-density matter, see also [Zdunik and Haensel (2013)] for the justification of its validity in the case of color superconducting quark matter EoS. The first demonstration that high-mass twin stars and thus a corresponding high-mass third family sequence with $M_{\mathrm{max}} > 2\ M_\odot$ were possible, has been given in [Alvarez-Castillo and Blaschke (2013)]. The intricacy of an equation of state describing a third family of stars (with twin stars at high or low masses as a consequence) consists in the fact that one needs, on the one hand, a sufficiently large jump in energy density $\Delta\varepsilon$ and a relatively low critical energy density ε_c at the transition point to fulfill the Seidov criterion (7.1) for gravitational instability (i.e., a stiff nuclear matter EoS has to be followed by a soft high-density one), while on the other the high-density EoS needs to become sufficiently stiff directly after the phase transition, without violating the causality condition ($c_s^2 < 1$). With the CSS parametrization, these constraints could be fulfilled relatively straightforwardly by dialing $c_s^2 = 1$ and adjusting a sufficiently large value of $\Delta\varepsilon$ by hand.

The question arose whether a hybrid star EoS describing a third family of compact stars with a maximum mass above $2\ M_\odot$ could also be obtained when applying the standard scheme of a two-phase approach based on a realistic nuclear matter EoS and a microscopically well-founded quark matter EoS, both joined, e.g., by a Maxwell construction. A positive answer was given already in 2013, when two examples of this kind were presented in [Blaschke *et al.* (2013a)], where the excluded-volume corrected nuclear EoS APR and DD2 were joined with a quark matter EoS based on the nonlocal NJL model approach [Blaschke *et al.* (2007); Benic *et al.* (2014)], augmented with a density-dependent repulsive vector mean field that was constructed by employing a thermodynamically consistent interpolation scheme introduced in [Blaschke *et al.* (2013b)]. Such an interpolation scheme, based on the nonlocal, color superconducting NJL model of [Blaschke *et al.* (2007)], but extended to address also a density-dependent bag pressure that facilitates a softening of the quark matter EoS in the vicinity of the deconfinement transition, has recently been developed in [Alvarez-Castillo *et al.* (2019)], guided by a relativistic density functional (RDF) approach to quark matter [Kaltenborn *et al.* (2017)] with an effective confinement mechanism according to the string-flip model [Horowitz *et al.* (1985); Ropke *et al.* (1986)]. This approach has been applied very successfully to describe a whole class of hybrid EoS with a third family branch fulfilling the modern compact star constraints [Ayriyan *et al.* (2018); Alvarez-Castillo *et al.* (2019)]. In the RDF approach to the string-flip model of quark matter an essential element is the ansatz for the nonlinear density functional resembling confinement and embodying the aspect of in-medium screening of the string-type confining interaction within an excluded volume mechanism. Another nonlinearity term in the density functional embodies the stiffening of quark matter at high densities in a similar way as it was obtained from 8-quark interactions in the NJL model [Benic (2014)] that were an essential part of the description of high-mass twins in [Benic *et al.* (2015)]. Both nonlinearities, due to (de)confinement and high-density stiffening, are mimicked in a rather flexible way by the two-fold interpolation scheme suggested in [Alvarez-Castillo *et al.* (2019)] which can be reinterpreted as a generalization of the nonlocal NJL model with chemical-potential-dependent parameters.

Having discussed the successful approaches to construct EoS with a strong phase transition which accounts for third family branches of compact stars and can be recognized observationally by the mass twin phenomenon, we would like to mention which ingredients are indispensable for obtaining this feature and which approaches have failed to obtain it. An excellent illustration of the various possibilities to join by interpolation hadronic and quark matter phases which themselves have different characteristics of stiffness, can be found in the recent review [Baym *et al.* (2018)]. However, despite being quite general, the case of the third family branch and mass-twin compact stars could not be captured! The reason can be found elucidated in [Alvarez-Castillo *et al.* (2019)], where it is demonstrated for a representative set of hadronic as well as quark matter EoS of varying degree of stiffness, that either a phase transition in the relevant domain of densities is entirely absent or results in a hybrid star branch that is directly connected to the hadronic branch which therefore does not form a third family of stars. The generated patterns are very similar to the results of [Orsaria *et al.* (2014)] which also uses the nonlocal NJL model for describing the quark phase of matter. The clue to obtaining third family sequences within microscopically motivated studies is a subtle softening followed by a stiffening of quark matter that can be realized employing the thermodynamically consistent interpolation procedure between different parameterizations of the same quark EoS (e.g., varying the vector–meson coupling strength) before applying a Maxwell, Gibbs, or pasta phase transition construction.

We have described here the state-of-the-art modeling of EoS with a strong phase transition that is based on microscopic models of high-density (quark) matter. Besides these, there is a large number of simple EoS parameterizations that are also in use for discussing third family sequences fulfilling the constraint of a high maximum mass of the order of 2 $M_\odot$. These are basically the classes of CSS based models and multi-polytrope approaches. Without attempting completeness, we would like to mention [Alford *et al.* (2015); Alford and Han (2016); Christian *et al.* (2018); Alford and Sedrakian (2017); Paschalidis *et al.* (2018); Christian *et al.* (2019); Montana *et al.* (2019); Han and Steiner (2019)] from the class of CSS models. A particularly interesting work is the extension by Alford and Sedrakian [Alford and Sedrakian (2017)], who demonstrated that also a fourth family of hybrid stars can be obtained and besides mass twins there are also possible mass triplets. The multi-polytrope EoS have been a workhorse for numerical relativity studies of astrophysical scenarios for many years. The approach to constrain the multi-polytrope parameters from observations of masses and radii as introduced by Read *et al.* [Read *et al.* (2009)] has then been developed further with great resonance in the community by Hebeler *et al.* (2010, 2013).

While in Read *et al.* (2009) one already finds a one-parameter set describing high-mass twin stars that have not yet become a matter of interest in the community, the third family branch in the multi-polytrope approaches has been mainly overlooked (see, e.g., Raithel *et al.* (2016); Miller *et al.* (2019)), but was digged up in [Alvarez-Castillo and Blaschke (2017)] where it was found that one should have at least a four-polytrope ansatz and suitably chosen densities for the matching of the polytrope pieces of the EoS, see also Annala *et al.* (2018); Paschalidis *et al.* (2018); Hanauske *et al.* (2018).

An important issue when discussing strong first-order phase transitions is the appearance of structures of finite size, like bubbles and droplets in the boiling/condensation transitions of the water–vapor transformations. In general, different shapes in the new phase are possible like spherical, cylindrical and planar structures, which have been dubbed "pasta structures". Their size and thermodynamical favorability depend on the surface tension between the sub-phases and the effects of the Coulomb interaction, including screening. The resulting pressure in the mixed phase is then no longer constant as in the Maxwell construction case, but also not as dramatically changing when the surface tension is neglected [Glendenning (1992)]. For details concerning the quark–hadron transition pasta phases under neutron star constraints see, e.g., [Na *et al.* (2012); Yasutake *et al.* (2014); M. Spinella *et al.* (2016)]. It is interesting to note that a simple one-parameter parabolic approximation of the pressure versus chemical potential dependence can give a satisfactory agreement with a full pasta phase calculation and that the single parameter can be directly related to the surface tension [Maslov *et al.* (2019)]. It could be demonstrated that the third family feature of an EoS with strong phase transition is rather robust against pasta phase effects, see Ayriyan *et al.* (2018).

In the present work, we will use the multi-polytrope approach to the EoS describing a third family of compact stars and also discuss the effect of mimicking the pasta structures in the mixed phase by a polynomial interpolation. The results shall not depend qualitatively on these simplifying assumptions but be of rather general nature and also applicable to more realistic types of EoS as discussed above.

7.4.1. *Multi-polytrope approach to the EoS*

In this section we present an EoS model that features a strong first-order phase transition from hadron to quark matter. They are labeled "ACB" following (Paschalidis *et al.*, 2018) and consist of a piecewise polytropic representation (Read *et al.*, 2009; Hebeler *et al.*, 2013; Raithel *et al.*, 2016) of the EoS at supersaturation densities $(n_1 < n < n_5 \gg n_0)$:

$$P(n) = \kappa_i (n/n_0)^{\Gamma_i}, \quad n_i < n < n_{i+1}, \quad i = 1,\dots,4, \quad (7.54)$$

where Γ_i is the polytropic index in each of the density regions labeled by $i = 1,\dots,4$. We fix Γ_1 such that a stiff nucleonic EoS provided in (Hebeler *et al.*, 2013) can be described. The second polytrope shall correspond to a first-order phase transition, therefore in this region the pressure must be constant, given by $P_c = \kappa_2$ ($\Gamma_2 = 0$). The remaining polytropes, in regions 3 and 4, that lie above the phase transition shall correspond to high-density matter, like stiff quark matter.

In order to compute the remaining thermodynamic variables of the EoS, we utilize the formulae given in the Appendix of (Zdunik *et al.*, 2006),

$$P(n) = n^2 \frac{d(\varepsilon(n)/n)}{dn}, \quad (7.55)$$

$$\varepsilon(n)/n = \int dn \, \frac{P(n)}{n^2} = \frac{1}{n_0^{\Gamma_i}} \int dn \, \kappa n^{\Gamma_i - 2} = \frac{1}{n_0^{\Gamma_i}} \frac{\kappa \, n^{\Gamma_i - 1}}{\Gamma_i - 1} + C, \quad (7.56)$$

$$\mu(n) = \frac{P(n) + \varepsilon(n)}{n} = \frac{1}{n_0^{\Gamma_i}} \frac{\kappa \, \Gamma_i}{\Gamma_i - 1} n^{\Gamma_i - 1} + m_0, \quad (7.57)$$

where we fix the integration constant C by the condition that $\varepsilon(n \to 0) = m_0 \, n$. Now we can invert the above expressions to obtain

$$n(\mu) = \left[n_0^{\Gamma_i} (\mu - m_0) \frac{\Gamma_i - 1}{\kappa \Gamma_i} \right]^{1/(\Gamma_i - 1)}, \quad (7.58)$$

so that the chemical-potential-dependent pressure for the polytrope EoS (7.54) can be written as

$$P(\mu) = \kappa \left[n_0^{\Gamma_i}(\mu - m_0) \frac{\Gamma_i - 1}{\kappa \Gamma_i} \right]^{\Gamma_i/(\Gamma_i - 1)}. \qquad (7.59)$$

The above form of the pressure (7.59) is suitable to perform a Maxwell construction of a first-order phase transition between the hadron and quark phases. The model parameters for the mass twin cases that we consider in this work are given in Table 7.3.

The ACB4 EoS features a first-order phase transition at a rather high nucleon number density value, $n_2 = 0.3174$ fm^{-3} that produces an instability in a $2.0 M_\odot$ neutron star, providing an example of the high-mass twins phenomenon. On the contrary, the ACB5 EoS presents a phase transition that occurs at the lower value of $n_2 = 0.284$ fm^{-3}. In that case the instability occurs for stars with $1.4 M_\odot$, providing a scenario for low-mass twin stars. This low density

Table 7.3   EoS models ACB4 and ACB5 (Paschalidis *et al.*, 2018). The parameters are defined in Eq. (7.54) in the main text. The first polytrope ($i = 1$) describes the nuclear EoS at supersaturation densities, the second polytrope ($i = 2$) corresponds to a first-order phase transition with a constant pressure P_c for densities between n_2 and n_3. The remaining polytropes lie in regions 3 and 4, i.e., above the phase transition and correspond to high-density matter, e.g., quark matter. The last column shows the maximum masses $M_{\max}$ on the hadronic (hybrid) branch corresponding to region 1 (4). The minimal mass $M_{\min}$ on the hybrid branch is shown for region 3.

ACB	i	Γ_i	κ_i [MeV/fm^3]	n_i [1/fm^3]	$m_{0,i}$ [MeV]	$M_{\max/\min}$ [$M_\odot$]
4	1	4.921	2.1680	0.1650	939.56	2.01
	2	0.0	63.178	0.3174	939.56	—
	3	4.000	0.5075	0.5344	1031.2	1.96
	4	2.800	3.2401	0.7500	958.55	2.11
5	1	4.777	2.1986	0.1650	939.56	1.40
	2	0.0	33.969	0.2838	939.56	—
	3	4.000	0.4373	0.4750	995.03	1.39
	4	2.800	2.7919	0.7500	932.48	2.00

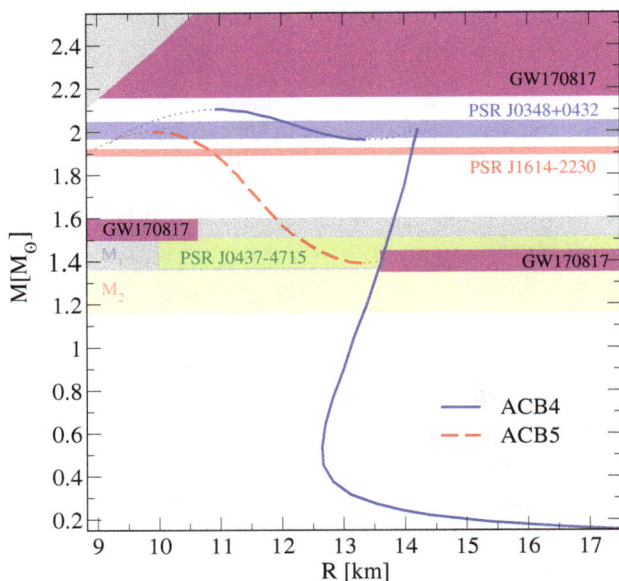

Fig. 7.1   Mass–radius relationship. The two curves correspond to the compact star sequences resulting after integration of the TOV equations for the ACB4 and ACB5 EoS. The blue, red and yellow regions correspond to mass measurements of the PSR J0348+0432, PSR J1614-2230 and PSR J0437-4715 pulsars, respectively. The latter is the target of the NICER detector that shall provide a measurement for its radius (Arzoumanian *et al.*, 2009). The areas labeled M1 and M2 correspond to the mass estimates of the compact stars that merged in the GW170817 event that was detected through gravitational radiation (Abbott *et al.*, 2017). The magenta marked areas labeled GW170817 are excluded by the GW170817 event (Bauswein *et al.*, 2017). An upper limit on the maximum mass of non-rotating compact stars of $2.16 M_\odot$ has been estimated in [Rezzolla *et al.* (2018)]. The region excluded by this estimate is shown by the magenta area, too. This limit will be reconsidered again later in light of the material presented in this work. The gray area in the upper left corner corresponds to a forbidden region where causality is violated.

value for the phase transition of about two times saturation density, is particularly feasible in neutron star matter, where the effect of the isospin asymmetry manifests in the so-called asymmetry energy which stiffens the EoS with respect to the symmetric case, equal number of protons and neutrons in hadronic matter. Figure 7.1 shows the resulting mass–radius curves for these two EoS featuring mass twins

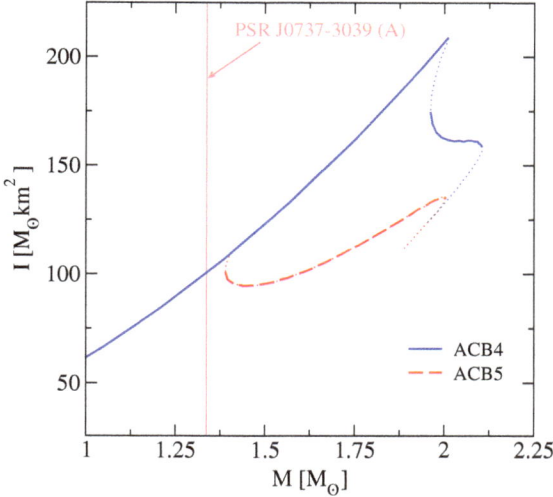

Fig. 7.2 Moment of inertia as a function of gravitational mass of the star for the two EoS cases ACB4 and ACB5. For an orientation, we indicate the precisely determined mass of the star PSR J0737-3039 (A), for which a measurement of the moment of inertia will become possible soon. This will provide further constraints on the EoS of dense matter and astrophysical scenarios involving compact stars.

together with measurements and constraint regions. Figure 7.2 shows moment of inertia as a function of gravitational mass of the star for the two EoS cases ACB4 and ACB5.

7.4.2. *EoS including mixed phase effects (pasta phases)*

In this section we introduce a mixed phase approach to mimic pasta structures in regions of both the hadronic and quark EoS around the Maxwell critical point (μ_c, P_c). The method used is the replacement interpolation method (RIM) (Abgaryan *et al.*, 2018) that consists of replacing the EoS in the aforementioned domain by a polynomial function:

$$P_M(\mu) = \sum_{q=1}^{N} \alpha_q (\mu - \mu_c)^q + (1 + \Delta_P) P_c, \qquad (7.60)$$

with Δ_P as free parameter that adds pressure to the mixed phase at μ_c. Generally, all parameterizations of the type shown in Eq. (7.60) for the mixed phase pressure are even order ($N = 2k$, $k = 1, 2, \ldots$) polynomials which we refer to as G_k. In order to smoothly match the EoS at μ_H and μ_Q up to the k-th derivative of the pressure the following conditions shall be fulfilled:

$$P_H(\mu_H) = P_M(\mu_H), \tag{7.61}$$

$$P_Q(\mu_Q) = P_M(\mu_Q), \tag{7.62}$$

$$\frac{\partial^k}{\partial \mu^k} P_H(\mu_H) = \frac{\partial^k}{\partial \mu^k} P_M(\mu_H), \tag{7.63}$$

$$\frac{\partial^k}{\partial \mu^k} P_Q(\mu_Q) = \frac{\partial^k}{\partial \mu^k} P_M(\mu_Q), \tag{7.64}$$

with α_q, μ_H and μ_Q being determined by the above system of equations.

For the sake of simplicity we employ the parabolic model G_1 of the RIA as introduced in (Ayriyan and Grigorian, 2018; Ayriyan *et al.*, 2018):

$$P_M(\mu) = \alpha_2(\mu - \mu_c)^2 + \alpha_1(\mu - \mu_c) + (1 + \Delta_P)P_c, \tag{7.65}$$

where the parameters α_1, α_2, μ_H and μ_Q are to be determined as described above, from the continuity conditions at the borders of the mixed phase:

$$P_H(\mu_H) = P_M(\mu_H), \tag{7.66}$$

$$P_Q(\mu_Q) = P_M(\mu_Q), \tag{7.67}$$

$$n_H(\mu_H) = n_M(\mu_H), \tag{7.68}$$

$$n_Q(\mu_Q) = n_M(\mu_Q). \tag{7.69}$$

Figure 7.3 shows a schematic representation of the RIM method based on the Maxwell construction between the hadronic and quark EoS.

In addition, Fig. 7.4 shows the mixed phase equations of state for both low and high mass twins. The effect of the mimicked geometrical

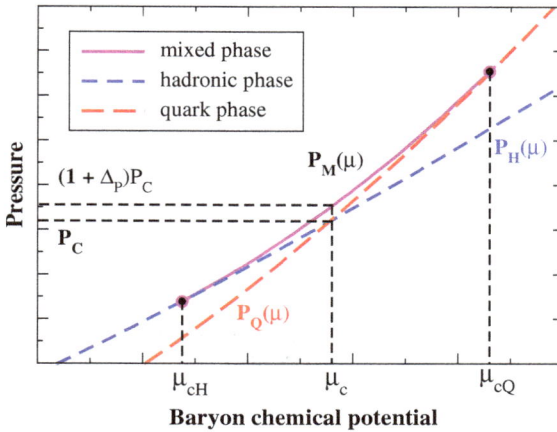

Fig. 7.3 Replacement interpolation function $P_M(\mu)$ obtained from the mixed phase constructions around the Maxwell construction. The resulting EoS connects the three points $P_H(\mu_H)$, $P_c + \Delta P = P_c(1 + \Delta_P)$, and $P_Q(\mu_Q)$. Figure taken from (Abgaryan *et al.*, 2018).

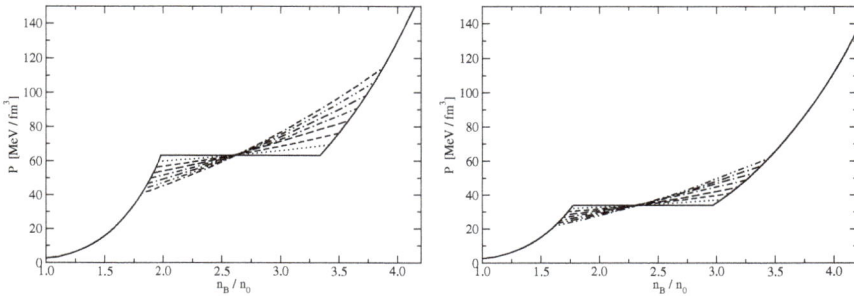

Fig. 7.4 Mixed phase mass twins equations of state for high mass NS onset (left) and low mass NS onset (right). The horizontal plateau at the phase transition corresponds to the Maxwell construction case. As the Δ_P parameter is increased successively, the plateau gives way to straight lines with increasing slope values.

structures is quantified by the Δ_P parameter. It is evident that the order of the G function will result in whether or not there are discontinuities for the derivatives of the P_M function. For instance, the square of the speed of sound, c_s^2, is proportional to the second derivative of G with respect to μ, see Fig. 7.5.

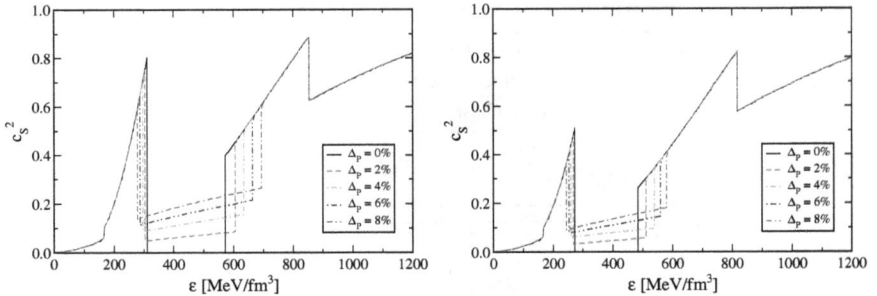

Fig. 7.5   Squared speed of sound as a function of the energy density for the mixed phase mass twins equations of state with high mass HS onset (ACB4, left panel) and low mass HS onset (ACB5, right panel).

The result is that G_1 presents a clear discontinuity in the speed of sound at ε_c and $\varepsilon_c + \Delta\varepsilon$, whereas in between, i.e., in the latent heat region, the speed of sound slightly increases. On the contrary, the construction G_2 allows for a continuous speed of sound, however it is not smoothly connected at ε_c and $\varepsilon_c + \Delta\varepsilon$. Only G_3 is capable of joining smoothly the speed of sound between the hadron and quark EoS at the critical points.

7.5. Results

7.5.1. *TOV solutions for mixed phase models*

In Fig. 7.6 we show the results of the mass–radius diagram as a solution of the TOV equations for the equations of state ACB4 (left panel) which exhibits high mass twin stars and ACB5 (right panel) which describes low mass twins, depending on the value of the mixed phase parameter Δ_P. In the insets we give a magnified view on the region of the maximum mass of the hadronic branch of the sequence, where the dotted lines indicate the unstable solutions that qualify the corresponding EoS as one with a third family. We can read off to the accuracy of the given 1% steps what the critical value for Δ_P

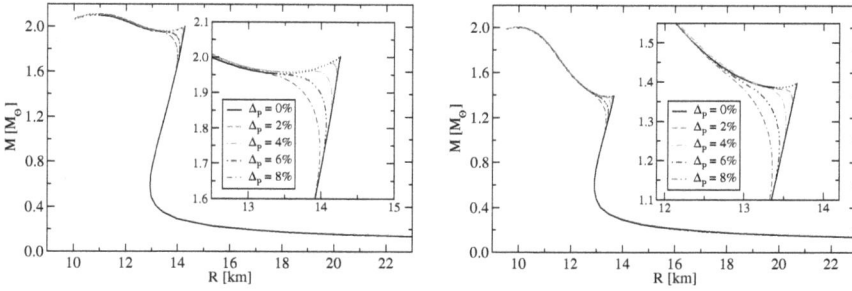

Fig. 7.6   Mass–radius diagram as a solution of the TOV equations for the equations of state ACB4 (left panel) which exhibits high mass twin stars and ACB5 (right panel) which describes low mass twins, depending on the value of the mixed phase parameter Δ_P.

is when the disconnected second and third families would merge to a connected hybrid star sequence.

While for the case of ACB4 the variation of Δ_P does not affect the mass–radius diagram in the mass region of the compact star merger GW170817, the corresponding variation for ACB5 leads to strong effects in that mass region. We therefore consider the tidal deformability in both cases in the next subsection.

7.5.2. *Tidal deformability predictions*

Together with the solution of the TOV equations, one can solve for the dimensionless tidal deformability $\Lambda(M)$ for the given EoS. After that, one can construct the corresponding lines in the $\Lambda_1 - \Lambda_2$ diagram of the binary compact star merger GW170817 for which the individual masses M_1 and M_2 of the two compact stars fulfill the constraint derived from the detected gravitational wave signal of the inspiral phase of the merger [Abbott *et al.* (2017)]. These lines can be overlaid to the constraint derived from the LVC observation, as shown in Fig. 7.7. As to be expected from Fig. 7.6, only in the case of ACB5 we can note an effect of the mixed phase construction while the results for ACB4 are inert against changes of the mixed

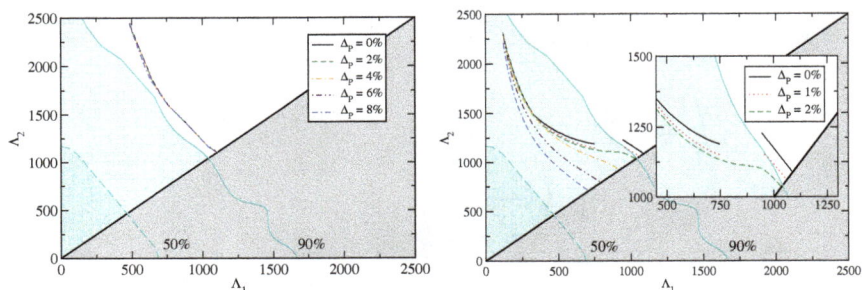

Fig. 7.7   Tidal deformability constraint from GW170817 for equations of state for high mass twins (left) and low mass twins (right).

phase parameter, because it influences the mass–radius diagram in a region of masses that is inaccessible to the gravitational wave signal of the inspiral phase and the effects of tidal deformation. Moreover, we notice that ACB4 is too stiff an EoS to fulfill the compactness constraint from GW170817. The EoS ACB5, however, with the early onset of the phase transition, becomes a soft EoS due to the mixed phase effects and for the largest values of the mixed phase parameter Δ_P is similar to a soft hadronic EoS despite the fact that the compact stars consist of extended regions of quark matter in pure or mixed phases.

In the following section, we will consider the effects of fast rotation on the sequences of hybrid star solutions and shall obtain a qualitative difference in the characteristics of pure phase (hadronic) and hybrid stars concerning their maximum masses which are relevant for the discussion of the phenomenology of binary compact star mergers and their implications for the state of superdense matter.

7.5.3. *Rotating compact star solutions*

In this subsection we present the numerical solutions for rotating hybrid star sequences in full GR equations for axial symmetry as obtained with the RNS code described in Sec. 7.2.2 and in the perturbative expansion up to order Ω^2 (slow rotation approximation) that was explained in Sec. 7.2.3. We relate these solutions to those

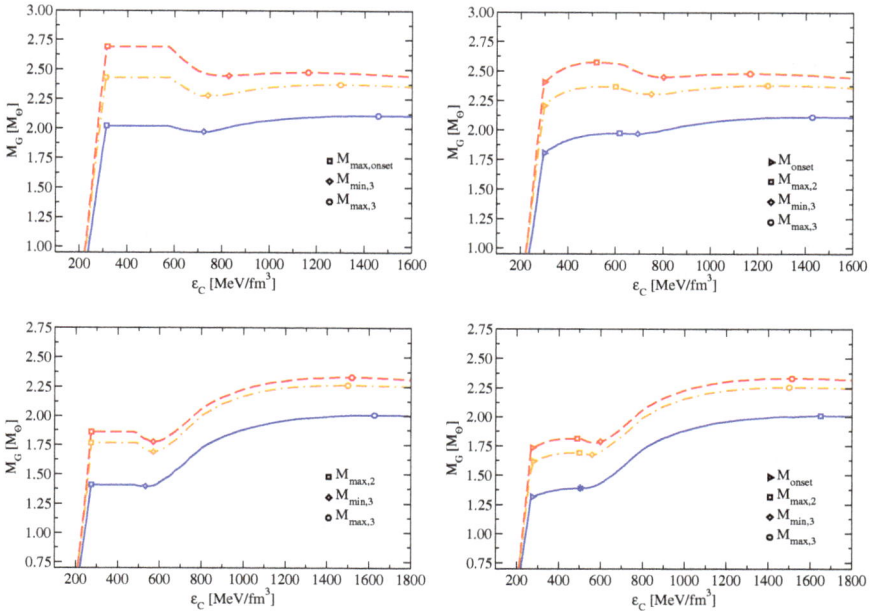

Fig. 7.8   The mass–central energy density diagram for compact star configurations for the EoS ACB4 (upper panels) and ACB5 (lower panels). The *left* panels are for $\Delta_P = 0$ (Maxwell construction) and the *right* ones are for the critical value of the mixed-phase parameter ($\Delta_P = 0.04$ for ACB4 and $\Delta_P = 0.02$ for ACB5) for which the third family vanishes in the static case because it joins the second family of neutron stars. Each panel shows three curves: the solution of the TOV equations for the static case (blue solid line), the solution of the slow rotation case (Ω^2 approximation) for rotation at the Kepler frequency Ω_K (orange dash-dotted line) and the full solution of the axisymmetric Einstein equations with the RNS code for $\Omega = \Omega_K$ (red dashed line). The symbols denote the onset mass for deconfinement (triangle right), the maximum mass on the $2^{\rm nd}$ family branch (square), the minimum mass (diamond) and the maximum mass (circle) on the $3^{\rm rd}$ family branch.

of the TOV equations for the static case of spherical symmetry discussed in Sec. 7.2.4.

In Fig. 7.8 we show the gravitational mass as a function of the central energy density for both multi-polytrope EoS, ACB4 (with a deconfinement phase transition at high mass, upper panels) and ACB5 (with a transition at the typical compact star mass of 1.4 $M_\odot$,

lower panels) for non-rotating (blue solid lines) and maximally rotating ($\Omega = \Omega_K$) stars in full GR (red dashed lines) and slow rotation
approximation (orange dash-dotted line). In the left panels for the
Maxwell construction case the jump in the central energy density by
about a factor two at the onset of the transition is clearly seen and
such an amount of latent heat is sufficient, according to the Seidov
criterion (7.1), to trigger a gravitational instability which occurs in
the region of densities where

$$\frac{dM}{d\varepsilon} < 0. \tag{7.70}$$

At the onset of this instability the gravitational mass of the star has
reached the maximum attainable on the second family of compact
stars, denoted by $M_{\text{max},2}$. For the Maxwell construction case, this
mass is degenerate with that for the onset of the phase transition,
M_{onset}. Due to the absence of a pressure gradient in the interval of
densities corresponding to the mixed phase, this phase is not realized
in compact stars in this case. The EoS with mass twin compact star
sequences are characterized by the fact that the instability criterion
(7.70) is fulfilled in a finite interval of densities which is then followed
by another stable, rising branch of sequences, the so-called third family of compact stars. This behavior defines two more characteristic
masses: $M_{\text{min},3}$ at the lower and $M_{\text{max},3}$ at the upper turning point,
see the left panels of Fig. 7.8.

For the mixed phase constructions depicted in the right panels of
Fig. 7.8, the pressures at the onset and the end of the mixed phase are
not identical and thus, due to the corresponding pressure gradient, a
mixed phase can be realized in the star and the degeneracy between
M_{onset} and $M_{\text{max},2}$ is lifted. On the other hand, by our choice of the
value of the mixed phase parameter Δ_P close to the limiting value
for which the second and the third families of compact stars would
get connected, the instability vanishes and thus $M_{\text{min},3}$ joins $M_{\text{max},2}$,
so that for a slightly larger value of Δ_P both these masses can no
longer be identified since only the second family survives which for
$M > M_{\text{onset}}$ consists of hybrid stars.

These four characteristic masses for a given mass twin compact star EoS are given in Table 7.4 for the static case obtained by solving the TOV equations. In the last column the absolute maximum of the mass–radius curve for the given EoS is listed. In Tables 7.5 and 7.6 these five characteristic masses are listed for the sequences of stars rotating at the Kepler frequency Ω_K which are obtained from solutions of the axisymmetric Einstein equations in the Ω^2 approximation and in full General Relativity, respectively.

Table 7.4 Five characteristic masses extracted from solutions of the Tolman–Oppenheimer–Volkoff equations (superscript "TOV") for sequences of static configurations with the EoS ACB4 and ACB5 for the Maxwell construction case ($\Delta_P = 0$) and for the mixed phase construction with the limiting value of Δ_P for which the second and the third family branches join. M_{onset} is the maximum mass of the purely hadronic second family branch at the onset of deconfinement, $M_{\mathrm{max},2}$ denotes the maximum mass reached at the end of the mixed phase, $M_{\mathrm{min},3}$ and $M_{\mathrm{max},3}$ are the minimum and the maximum mass on the third family branch of the sequence. The maximum mass of the whole sequence for a given EoS is denoted as M_{max}.

EoS	Δ_P	$M_{\mathrm{onset}}^{\mathrm{TOV}}$	$M_{\mathrm{max},2}^{\mathrm{TOV}}$	$M_{\mathrm{min},3}^{\mathrm{TOV}}$	$M_{\mathrm{max},3}^{\mathrm{TOV}}$	$M_{\mathrm{max}}^{\mathrm{TOV}}$
2^*ACB4	0%	2.020	2.020	1.969	2.107	2.107
	4%	1.801	1.970	1.965	2.108	2.108
2^*ACB5	0%	1.404	1.404	1.393	2.004	2.004
	2%	1.312	1.386*	1.386*	2.006*	2.006

Table 7.5 Same as Table 7.4, but now for solutions of the axisymmetric Einstein equations in the slow rotation approximation, the perturbative expansion to order Ω^2 denoted by the corresponding superscript.

EoS	Δ_P	$M_{\mathrm{onset}}^{\Omega^2}$	$M_{\mathrm{max},2}^{\Omega^2}$	$M_{\mathrm{min},3}^{\Omega^2}$	$M_{\mathrm{max},3}^{\Omega^2}$	$M_{\mathrm{max}}^{\Omega^2}$
2^*ACB4	0%	2.426	2.426	2.274	2.369	2.426
	4%	2.200	2.362	2.301	2.373	2.373
2^*ACB5	0%	1.759	1.759	1.685	2.261	2.261
	2%	1.611	1.685	1.670	2.251	2.251

Table 7.6   Same as Table 7.4, but now for solutions of the full system of Einstein equations for uniform rotation in axial symmetry [Cook *et al.* (1994)] using the RNS code. The corresponding results are denoted by the superscript "rot".

EoS	Δ_P	$M_{\mathrm{onset}}^{\mathrm{rot}}$	$M_{\mathrm{max},2}^{\mathrm{rot}}$	$M_{\mathrm{min},3}^{\mathrm{rot}}$	$M_{\mathrm{max},3}^{\mathrm{rot}}$	$M_{\mathrm{max}}^{\mathrm{rot}}$
2*ACB4	0%	2.686	2.686	2.442	2.472	2.686
	4%	2.401	2.569	2.445	2.475	2.569
2*ACB5	0%	1.855	1.855	1.770	2.328	2.328
	2%	1.727	1.807	1.780	2.328	2.328

Inspecting the rotating star sequences for the close-to-critical mixed phase parameter Δ_P, we observe that due to the rotation the star branches with mixed phase and pure quark matter core can get disconnected so that the phenomenon of a third family reappears. *Vice versa*, upon spin-down from a supramassive star configuration at maximal rotation frequency (created, e.g., in a binary neutron star merger) which is stable on the hadronic or mixed phase branch, it may end up either as a black hole or on the hybrid star branch for such mixed-phase EoS. The scenario of a delayed collapse to a black hole is of special importance for interpreting GW170817 and will therefore be discussed below in further detail. In this context appears the question whether between the above-introduced characteristic masses at maximal and at zero rotation frequency hold EoS-independent, so-called *universal relations* that have been investigated for the maximum mass of hadronic EoS [Bozzola *et al.* (2018); Rezzolla *et al.* (2018)] and recently also for hybrid EoS including the Maxwell construction cases of ACB4 and ACB5 [Bozzola *et al.* (2019)]. We shall come back to this issue below.

Here we like to remark that the maximum masses on the second and third family branches correspond to stars with very different central (energy) densities. This may be the clue to understanding the fact that the increase in mass for stars on the more compact third family branch is smaller than for stars on the second family one because of their smaller radii and thus smaller moment of inertia (7.53) and rotational energy, see the gravitational mass vs.

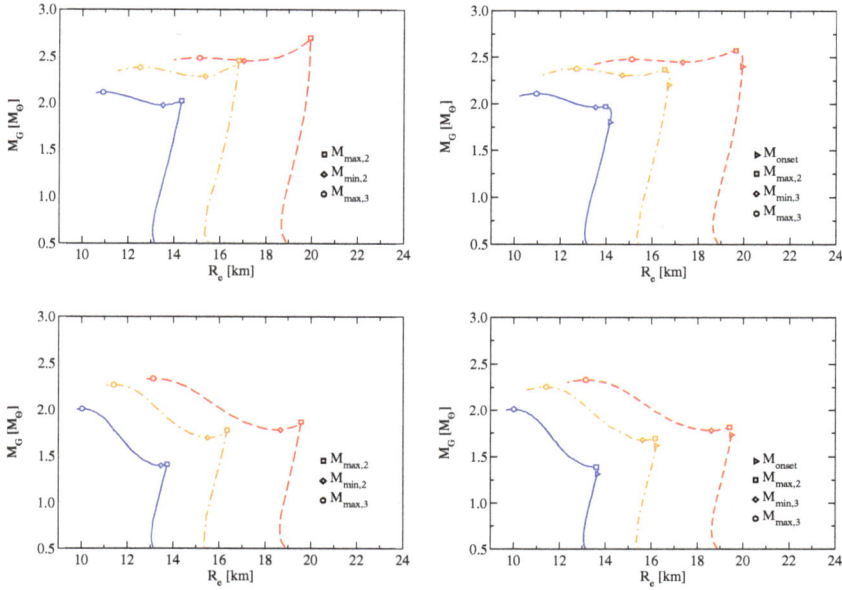

Fig. 7.9   The mass–radius diagram for compact star configurations for the EoS ACB4 (upper panels) and ACB5 (lower panels). The *left* panels are for $\Delta_P = 0$ (Maxwell construction) and the *right* ones are for the critical value of the mixed-phase parameter ($\Delta_P = 0.04$ for ACB4 and $\Delta_P = 0.02$ for ACB5) for which the third family vanishes in the static case because it joins the second family of neutron stars. Each panel shows three curves: the solution of the TOV equations for the static case (blue solid line), the solution of the slow rotation case (Ω^2 approximation) for rotation at the Kepler frequency Ω_K (orange dash-dotted line) and the full solution of the axisymmetric Einstein equations with the RNS code for $\Omega = \Omega_K$ (red dashed line). The symbols denote the onset mass for deconfinement (triangle right), the maximum mass on the second family branch (square), the minimum mass (diamond) and the maximum mass (circle) on the third family branch.

equatorial radius in Fig. 7.9. For a more quantitative discussion of the effects of rotation on the masses of the sequences, we extract from the Tables 7.4–7.6 the ratios of the characteristic masses on the rotating sequences to those on the static ones in Tables 7.7 and 7.8 for the Ω^2 approximation and the full GR solution, respectively. For completeness, we give the ratio of the characteristic masses between the two rotation solutions in Table 7.9.

Table 7.7   Ratios formed by the five characteristic masses at maximal rotation frequency Ω_K in the slow rotation approximation (superscript "Ω^2") relative to the static case (superscript "TOV") calculated with the EoS ACB4 and ACB5 for the Maxwell construction case ($\Delta_P = 0$) and for the mixed phase construction with the limiting value of Δ_P.

EoS	Δ_P	$\dfrac{M_{\mathrm{onset}}^{\Omega^2}}{M_{\mathrm{onset}}^{\mathrm{TOV}}}$	$\dfrac{M_{\mathrm{max},2}^{\Omega^2}}{M_{\mathrm{max},2}^{\mathrm{TOV}}}$	$\dfrac{M_{\mathrm{min},3}^{\Omega^2}}{M_{\mathrm{min},3}^{\mathrm{TOV}}}$	$\dfrac{M_{\mathrm{max},3}^{\Omega^2}}{M_{\mathrm{max},3}^{\mathrm{TOV}}}$	$\dfrac{M_{\mathrm{max}}^{\Omega^2}}{M_{\mathrm{max}}^{\mathrm{TOV}}}$
ACB4	0%	1.201	1.201	1.155	1.124	1.151
	4%	1.222	1.199	1.171	1.126	1.126
ACB5	0%	1.253	1.253	1.210	1.128	1.128
	2%	1.228	1.216*	1.205*	1.122*	1.122

Table 7.8   Same as Table 7.7, but now for the full solutions of the axisymmetric Einstein equations with the RNS code relative to the solutions of the TOV equations for the static case.

EoS	Δ_P	$\dfrac{M_{\mathrm{onset}}^{\mathrm{rot}}}{M_{\mathrm{onset}}^{\mathrm{TOV}}}$	$\dfrac{M_{\mathrm{max},2}^{\mathrm{rot}}}{M_{\mathrm{max},2}^{\mathrm{TOV}}}$	$\dfrac{M_{\mathrm{min},3}^{\mathrm{rot}}}{M_{\mathrm{min},3}^{\mathrm{TOV}}}$	$\dfrac{M_{\mathrm{max},3}^{\mathrm{rot}}}{M_{\mathrm{max},3}^{\mathrm{TOV}}}$	$\dfrac{M_{\mathrm{max}}^{\mathrm{rot}}}{M_{\mathrm{max}}^{\mathrm{TOV}}}$
ACB4	0%	1.330	1.330	1.240	1.173	1.275
	4%	1.333	1.304	1.244	1.174	1.174
ACB5	0%	1.321	1.321	1.270	1.162	1.162
	2%	1.316	1.304*	1.284*	1.161*	1.161

Table 7.9   Same as Table 7.7, but now the full solutions of the axisymmetric Einstein equations (superscript "RNS") are related to those in the slow rotation approximation (superscript "Ω^2").

EoS	Δ_P	$\dfrac{M_{\mathrm{onset}}^{\mathrm{RNS}}}{M_{\mathrm{onset}}^{\Omega^2}}$	$\dfrac{M_{\mathrm{max},2}^{\mathrm{RNS}}}{M_{\mathrm{max},2}^{\Omega^2}}$	$\dfrac{M_{\mathrm{min},3}^{\mathrm{RNS}}}{M_{\mathrm{min},3}^{\Omega^2}}$	$\dfrac{M_{\mathrm{max},3}^{\mathrm{RNS}}}{M_{\mathrm{max},3}^{\Omega^2}}$	$\dfrac{M_{\mathrm{max}}^{\mathrm{RNS}}}{M_{\mathrm{max}}^{\Omega^2}}$
ACB4	0%	1.107	1.107	1.074	1.043	1.107
	4%	1.091	1.087	1.063	1.043	1.082
ACB5	0%	1.054	1.054	1.050	1.030	1.030
	2%	1.072	1.072	1.065	1.034	1.034

7.6. Implications for the Phenomenology of Compact Stars

We investigate the consequences of a strong phase transition in the EoS for dense compact star matter for sequences of configurations in the mass–radius as well as mass–central (energy) density plane, with and without rotation. While for isolated pulsars even the highest known spin frequencies are well below the Kepler frequency so that no strong modification of the TOV solution occurs. In the era of multi-messenger astronomy, with compact star mergers being accessible to detection by their GW signal from the inspiral phase and soon also from the post-merger state, the "übermassive" [Espino and Pascha-lidis (2019)] as well as supramassive star solutions play a role in the interpretation of the observations. While the former are solutions for differentially rotating configurations at the mass shedding limit, the latter are uniformly rotating objects with a frequency close to the Kepler one [Shibata *et al.* (2019)]. Recently, for this case, EoS inde-pendent, so-called "universal" relationships have been derived which relate the maximum mass of the supramassive star sequence to the static one from the solution of the TOV equation. Such relationships, once confirmed, are particularly useful in order to make general pre-dictions or draw conclusions from merger phenomenology for con-straints limiting the EoS properties. In this context we would like to mention the upper limit on the maximum mass of neutron stars that has been extracted from the GW signal and phenomenology of GW170817 [Shibata *et al.* (2017); Margalit and Metzger (2017); Rezzolla *et al.* (2018)].

It has been known for a long time [Haensel *et al.* (2007)] that there is a relationship between the maximum mass of uniformly rotating cold neutron stars at the maximum frequency and the maximum mass of the TOV equation solution for static stars, given by

$$M_{\max}(\Omega_K) = \alpha M_{\max}^{\mathrm{TOV}}, \tag{7.71}$$

where recently in [Breu and Rezzolla (2016)] the universality of this relationship was confirmed for a very large set of hadronic EoS (with-out a deconfinement phase transition) with the coefficient $\alpha = 1.20$.

The hypothesis that the relation (7.71) can be extended to include hybrid stars with a strong phase transition and even with third family sequences has recently been investigated in [Bozzola *et al.* (2019)] and following the argumentation of [Ruiz *et al.* (2018); Most *et al.* (2018); Rezzolla *et al.* (2018)] it leads to a limitation for the maximum mass of static neutron stars as

$$2.07\, M_\odot \lesssim M_{\max}^{\mathrm{TOV}} \lesssim 2.23\, M_\odot. \qquad (7.72)$$

The value of $M_{\max} = 2.591\, M_\odot$ was extracted for the core mass of the compact star merger GW170817 in [Rezzolla *et al.* (2018)]. We confirm the finding of [Bozzola *et al.* (2019)] that the coefficient spans a range of values for which we find $0.16 < \alpha < 0.33$, see Table 7.8. The lower limit in Eq. (7.72) comes from the new high-mass pulsar PSR J0740+6620 for which a mass $2.14^{+0.10}_{-0.09}\, M_\odot$ has been determined by Cromartie *et al.* (2019) by measuring the Shapiro delay, and the upper limit in our case is $2.23\, M_\odot$ for the lowest value of $\alpha = 1.16$. In order to not come in conflict with the pulsar mass measurement of Cromartie *et al.* (2019), there is an upper limit for the admissible value of $\alpha = 1.25$, corresponding to the lower limit at the 1σ level of the PSR J0740+6620 mass, $M_{\max}^{\mathrm{TOV}} = 2.07\, M_\odot$. Taking

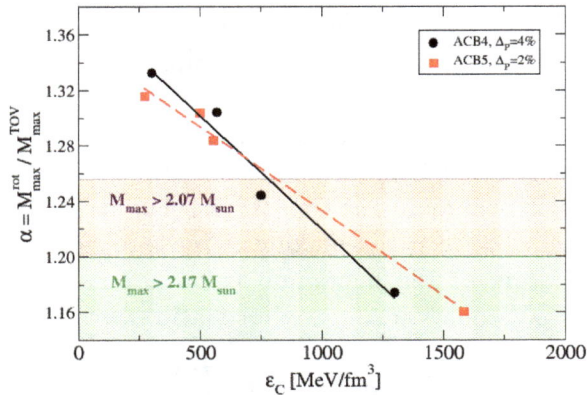

Fig. 7.10   Ratio of the mass for maximally rotating stars to that of a static star as a function of the central energy density. For a comparison the value 1.20 would be the maximum value compatible with a lower limit on the maximum mass of (non-rotating) pulsars of $2.17\, M_\odot$ [Cromartie *et al.* (2019)].

the central value, $M_{\max}^{\text{TOV}} = 2.17\ M_\odot$, would correspond to $\alpha = 1.20$, see Fig. 7.10. This figure illustrates one of the main findings of this contribution, the dependence of the coefficient α in Eq. (7.71) on the central (energy) density of the stellar configuration that can be fitted by a linear regression to the values we determined at the positions corresponding to the characteristic masses to be

$$\alpha = a - b\varepsilon_c, \qquad (7.73)$$

where $a_{ACB4-4} = 1.38 \pm 0.07$, $b_{ACB4-4} = 0.12 \pm 0.01\ \text{fm}^3/\text{GeV}$ and $a_{ACB5-2} = 1.37 \pm 0.07$, $b_{ACB5-2} = 0.16 \pm 0.02\ \text{fm}^3/\text{GeV}$. We may conclude that only those states of matter are allowed for the inner core of a compact star at maximum mass of $2.07\ M_\odot$ ($2.17\ M_\odot$) which belong to a high-density region with $\varepsilon \geq 0.78\ \text{GeV}/\text{fm}^3$ ($\varepsilon \geq 1.12\ \text{GeV}/\text{fm}^3$).

Anyway, the main effect of the strong phase transition is a higher compactness of the high mass stars than in the purely hadronic case which moreover goes along with a smaller mass increase due to maximal rotation which entails the increase of the upper limit for the maximum mass relative to the purely hadronic case discussed in [Shibata *et al.* (2017); Margalit and Metzger (2017); Rezzolla *et al.* (2018)]. Thus the high-density phase transition removes a certain tension from the discussion of the upper limit for the maximum mass and could be used as an argument in favor of the suggestion that a strong phase transition actually takes place in compact stars!

7.7. Summary and Conclusions

Stimulated by the unprecedented progress in observational astronomy, compact stars have become superb astrophysical laboratories for a broad range of physical studies. This is particularly the case for neutron stars, since their observations carry information about the fundamental building blocks of matter and even of the fabric of space itself. Against this background we present a systematic investigation of the properties of compact stellar mass twins (i.e., the so-called third family of compact stars), which, according to theory, may exist in the mass–radius region between neutron stars and stellar-mass black holes. Particular emphasis is given to modeling the rotational properties of compact mass twins for multi-polytrope models for the equation of state of ultra-dense stellar matter that fulfill the constraint established for the maximum mass of a neutron star. The main results of our investigation can be summarised as follows:

1) The existence of mass twins invariably signals the existence of a strong phase transition in ultra-dense matter. The extreme softening of the equation of state caused by the strong phase transition increases the gravitational field so much that the star becomes gravitationally unstable over a certain range of densities, where just a certain fraction of the star's core is in the new high-density phase. Eventually the star becomes stable again when about half of the matter in its core is in the new phase of matter. The new star has gravitational mass that is less than the maximal mass of the hadronic stellar sequence.

   The observation of mass twins would indicate a strong phase transition so that from the existence of a strong first-order phase transition in one corner of the QCD phase diagram and a crossover behavior in another, one could conclude that at least one critical endpoint (CEP) must exist. This would be very reassuring for large scale experimental heavy-ion collision programs set up for the search for the CEP.

2) The mixed phase construction which mimics the pasta phase is in accordance with a full pasta calculation. The result is a "smearing" of the phase transition over a certain pressure region, which is similar to the Gibbs construction in matter with more than one conserved charge and where charge conservation need not be fulfilled locally but rather globally [Glendenning (1992)]. This construction makes not only the approach to the strong phase transition more realistic, but has also great advantages for the numerical realization of phenomenological scenarios of the phase transition in rapidly rotating stars studied in numerical relativity.

3) Lastly, we address the conjecture of an upper limit on the maximum masses of non-rotating compact stars from the

phenomenology of GW170817 and its associated kilonova event. The conjecture is based on a universal relation between the maximum masses of uniformly rotating stars and that of static TOV solutions for the same EoS, which was demonstrated to hold for neutron star EoS without a phase transition [Breu and Rezzolla (2016)]. The stellar mass at which the high-density phase transition (such as deconfinement of quarks) sets in is currently unknown. But if the transition would occur below the maximum mass of the TOV solution, the quasi-universal relation will have to be revisited and the conclusions for the upper limit on the maximum mass have to be revised. We have found a quantitative criterion for the minimal central energy density in the maximum-mass configuration of a compact star that would correspond to the core of GW170817 after dynamical mass ejection [Rezzolla *et al.* (2018)]. Thus the EoS at high densities must be effectively soft, either in the form of a relatively soft hadronic EoS or as a hybrid EoS with a phase transition, since a hadronic EoS too stiff would lead to heavy hadronic stars too dilute to fulfill the constraint derived by us.

With these prospects for strong gravity and strong phase transitions in compact stars, we are looking forward to the next series of exciting discoveries in the just opened era of multi-messenger astronomy.

Acknowledgments

We would like to thank Cesar Zen Vasconcellos for the invitation to contribute to this book and for his patience and numerous reminders which encouraged us to complete the writeup despite interfering obligations. We are grateful to Luciano Rezzolla for his comments on the universality of the relation (7.71) which holds for hadronic EoS not only at the Keplerian limit. The work of D.B., A.A., H.G. has been supported by the Russian Science Foundation under grant 17-12-01427; F.W. acknowledges support by the U.S. National Science Foundation under Grant PHY-1714068. The authors are grateful to the European COST Actions CA15213 "THOR" and CA16214 "PHAROS" for supporting their networking activities.

References

Abbott, B. *et al.*, *Phys. Rev. Lett.* **119**(16) 161101, doi:10.1103/PhysRevLett.119. 161101, arXiv:1710.05832 [gr-qc] (2017).

Abgaryan, V., Alvarez-Castillo, D., Ayriyan, A., Blaschke, D., and Grigorian, H., *Universe* **4**(9), 94, doi:10.3390/universe4090094, arXiv:1807.08034 [astro-ph.HE] (2018).

Alford, M.G., Burgio, G.F., Han, S., Taranto, G., and Zappala, D., *Phys. Rev. D* **92**(8), 083002, doi:10.1103/PhysRevD.92.083002, arXiv:1501.07902 [nucl-th] (2015).

Alford, M.G. and Han, S., *Eur. Phys. J. A* **52**(3), 62, doi:10.1140/epja/ i2016-16062-9, arXiv:1508.01261 [nucl-th] (2016).

Alford, M.G., Han, S., and Prakash, M., *Phys. Rev. D* **88**(8), 083013, doi:10. 1103/PhysRevD.88.083013, arXiv:1302.4732 [astro-ph.SR] (2013).

Alford, M.G. and Sedrakian, A., *Phys. Rev. Lett.* **119**(16), 161104, doi:10.1103/ PhysRevLett.119.161104, arXiv:1706.01592 [astro-ph.HE] (2017).

Alvarez-Castillo, D.E., Bejger, M., Blaschke, D., Haensel, P., and Zdunik, L., arXiv:1506.08645 [astro-ph.HE] (2015).

Alvarez-Castillo, D.E. and Blaschke, D., Proving the CEP with compact stars? in *Proceedings, 17th Conference of Young Scientists and Specialists (AYSS '13)*, Dubna, Russia, April 8–12, 2013, arXiv:1304.7758 [astro-ph.HE] (2013).

Alvarez-Castillo, D.E. and Blaschke, D.B. *Phys. Rev. C* **96**(4), 045809, doi:10. 1103/PhysRevC.96.045809, arXiv:1703.02681 [nucl-th] (2017).

Alvarez-Castillo, D.E., Blaschke, D.B., Grunfeld, A.G., and Pagura, V.P., *Phys. Rev. D* **99**(6), 063010, doi:10.1103/PhysRevD.99.063010, arXiv:1805.04105 [hep-ph] (2019).

Annala, E., Gorda, T., Kurkela, A., and Vuorinen, A., *Phys. Rev. Lett.* **120**(17), 172703, doi:10.1103/PhysRevLett.120.172703, arXiv:1711.02644 [astro-ph.HE] (2018).

Arzoumanian, Z. *et al.*, arXiv:0902.3264 [astro-ph.HE] (2009).

Ayriyan, A., Bastian, N.U., Blaschke, D., Grigorian, H., Maslov, K., and Voskresensky, D.N., *Phys. Rev. C* **97**(4), 045802, doi:10.1103/PhysRevC.97. 045802, arXiv:1711.03926 [nucl-th] (2018).

Ayriyan, A. and Grigorian, H., *Eur. Phys. J. Web Conf.* **173**, 03003, doi:10.1051/ epjconf/201817303003, arXiv:1710.05637 [astro-ph.HE] (2018).

Ayvazyan, N.S., Colucci, G., Rischke, D.H., and Sedrakian, A., *Astron. Astrophys.* **559**, A118, doi:10.1051/0004-6361/201322484, arXiv:1308.3053 [astro-ph.SR] (2013).

Bauswein, A., Just, O., Janka, H.-T., and Stergioulas, N., *Astrophys. J.* **850**(2), L34, doi:10.3847/2041-8213/aa9994, arXiv:1710.06843 [astro-ph. HE] (2017).

Baym, G., Hatsuda, T., Kojo, T., Powell, P.D., Song, Y., and Takatsuka, T., *Rept. Prog. Phys.* **81**(5), 056902, doi:10.1088/1361-6633/aaae14, arXiv:1707.04966 [astro-ph.HE] (2018).

Bejger, M., Blaschke, D., Haensel, P., Zdunik, J.L., and Fortin, M., *Astron. Astrophys.* **600**, A39, doi:10.1051/0004-6361/201629580, arXiv:1608.07049 [astro-ph.HE] (2017).

Benic, S., *Eur. Phys. J. A* **50**, 111, doi:10.1140/epja/i2014-14111-1, arXiv:1401.5380 [nucl-th] (2014).

Benic, S., Blaschke, D., Alvarez-Castillo, D.E., Fischer, T., and Typel, S., *Astron. Astrophys.* **577**, A40, doi:10.1051/0004-6361/201425318, arXiv:1411.2856 [astro-ph.HE] (2015).

Benic, S., Blaschke, D., Contrera, G.A., and Horvatic, D., *Phys. Rev. D* **89**(1), 016007, doi:10.1103/PhysRevD.89.016007, arXiv:1306.0588 [hep-ph] (2014).

Binnington, T. and Poisson, E., *Phys. Rev. D* **80**, 084018, doi:10.1103/PhysRevD. 80.084018, arXiv:0906.1366 [gr-qc] (2009).

Blaschke, D., Alvarez-Castillo, D.E., and Benic, S., *Proc. Sci.* **CPOD2013**, 063, doi:10.22323/1.185.0063, arXiv:1310.3803 [nucl-th] (2013a).

Blaschke, D., Alvarez Castillo, D.E., Benic, S., Contrera, G., and Lastowiecki, R., *Proc. Sci.* **171**, doi:10.22323/1.171.0249, arXiv:1302.6275 [hep-ph] (2013b).

Blaschke, D.B., Gomez Dumm, D., Grunfeld, A.G., Klahn, T., and Scoccola, N.N., *Phys. Rev. C* **75**, 065804, doi:10.1103/PhysRevC.75.065804, arXiv:nucl-th/0703088 [nucl-th] (2007).

Bozzola, G., Espino, P.L., Lewin, C.D., and Paschalidis, V., arXiv:1905.00028 [astro-ph.HE] (2019).

Bozzola, G., Stergioulas, N., and Bauswein, A., *Mon. Not. Roy. Astron. Soc.* **474**(3), 3557–3564, doi:10.1093/mnras/stx3002, arXiv:1709.02787 [gr-qc] (2018).

Breu, C. and Rezzolla, L., *Mon. Not. Roy. Astron. Soc.* **459**(1), 646, doi:10.1093/mnras/stw575, arXiv:1601.06083 [gr-qc] (2016).

Christian, J.-E., Zacchi, A., and Schaffner-Bielich, J., *Eur. Phys. J. A* **54**(2), 28, doi:10.1140/epja/i2018-12472-y, arXiv:1707.07524 [astro-ph.HE] (2018).

Christian, J.-E., Zacchi, A., and Schaffner-Bielich, J., *Phys. Rev. D* **99**(2), 023009, doi:10.1103/PhysRevD.99.023009, arXiv:1809.03333 [astro-ph.HE] (2019).

Chubarian, E., Grigorian, H., Poghosyan, G.S., and Blaschke, D., *Astron. Astrophys.* **357**, 968, arXiv:astro-ph/9903489 [astro-ph] (2000).

Cook, G.B., Shapiro, S.L., and Teukolsky, S.A., *Astrophys. J.* **422**, 227 (1994).

Cromartie, H.T. *et al.*, *Nat. Astron.* **3**, 439, arXiv:1904.06759 [astro-ph.HE] (2019).

Damour, T. and Nagar, A., *Phys. Rev. D* **80**, 084035, doi:10.1103/PhysRevD.80. 084035, arXiv:0906.0096 [gr-qc] (2009).

Espino, P. and Paschalidis, V., *Phys. Rev. D* **99**(8), 083017, doi:10.1103/PhysRevD.99.083017, arXiv:1901.05479 [astro-ph.HE] (2019).

Gerlach, U.H., *Phys. Rev.* **172**, 1325, doi:10.1103/PhysRev.172.1325 (1968).

Glendenning, N.K., *Phys. Rev. D* **46**, 1274, doi:10.1103/PhysRevD.46.1274 (1992).

Glendenning, N.K. and Kettner, C., *Astron. Astrophys.* **353**, L9, arXiv:astro-ph/9807155 [astro-ph] (2000).

Glendenning, N.K., *Compact Stars: Nuclear Physics, Particle Physics, and General Relativity* (Springer, 2000).

Haensel, P., Potekhin, A.Y., and Yakovlev, D.G. *Astrophys. Space Sci. Libr.* **326**, 1, doi:10.1007/978-0-387-47301-7 (2007).

Han, S. and Steiner, A.W., *Phys. Rev. D* **99**(8), 083014, doi:10.1103/PhysRevD.99.083014, arXiv:1810.10967 [nucl-th] (2019).

Hanauske, M., Yilmaz, Z.S., Mitropoulos, C., Rezzolla, L., and Stöcker, H., *Eur. Phys. J. Web Conf.* **171**, 20004, doi:10.1051/epjconf/201817120004 (2018).

Hartle, J.B., *Astrophys. J.* **150**, 1005, doi:10.1086/149400 (1967).

Hartle, J.B. and Thorne, K.S., *Astrophys. J.* **153**, 807, doi:10.1086/149707 (1968).

Haskell, B. and Melatos, A., *Int. J. Mod. Phys. D* **24**(3), 1530008, doi:10.1142/S0218271815300086, arXiv:1502.07062 [astro-ph.SR] (2015).

Hebeler, K., Lattimer, J.M., Pethick, C.J., and Schwenk, A., *Phys. Rev. Lett.* **105**, 161102, doi:10.1103/PhysRevLett.105.161102, arXiv:1007.1746 [nucl-th] (2010).

Hebeler, K., Lattimer, J.M., Pethick, C.J., and Schwenk, A., *Astrophys. J.* **773**, 11, doi:10.1088/0004-637X/773/1/11, arXiv:1303.4662 [astro-ph.SR] (2013).

Hinderer, T., *Astrophys. J.* **677**, 1216, doi:10.1086/533487, arXiv:0711.2420 [astro-ph] (2008).

Hinderer, T., Lackey, B.D., Lang, R.N., and Read, J.S., *Phys. Rev. D* **81**, 123016, doi:10.1103/PhysRevD.81.123016, arXiv:0911.3535 [astro-ph.HE] (2010).

Horowitz, C.J., Moniz, E.J., and Negele, J.W., *Phys. Rev. D* **31**, 1689, doi:10.1103/PhysRevD.31.1689 (1985).

Kaltenborn, M.A.R., Bastian, N.-U.F., and Blaschke, D.B., *Phys. Rev. D* **96**(5), 056024, doi:10.1103/PhysRevD.96.056024, arXiv:1701.04400 [astro-ph.HE] (2017).

Komatsu, H., Eriguchi, Y., and Hachisu, I., *Mon. Not. Roy. Astron. Soc.* **237**, 355 (1989).

Laarakkers, W.G. and Poisson, E., *Astrophys. J.* **512**, 282, doi:10.1086/306732, arXiv:gr-qc/9709033 [gr-qc] (1999).

Lattimer, J.M. and Prakash, M., *Phys. Rept.* **442**, 109, doi:10.1016/j.physrep.2007.02.003, arXiv:astro-ph/0612440 [astro-ph] (2007).

Lindblom, L., *Phys. Rev. D* **58**, 024008, doi:10.1103/PhysRevD.58.024008, arXiv:gr-qc/9802072 [gr-qc] (1998).

Margalit, B. and Metzger, B.D., *Astrophys. J.* **850**(2), L19, doi:10.3847/2041-8213/aa991c, arXiv:1710.05938 [astro-ph.HE] (2017).

Maslov, K., Yasutake, N., Ayriyan, A., Blaschke, D., Grigorian, H., Maruyama, T., Tatsumi, T., and Voskresensky, D.N., *Phys. Rev. C* **100**(2), 025802, doi:10.1103/PhysRevC.100.025802, arXiv:1812.11889 [nucl-th] (2019).

Miller, M.C., Chirenti, C., and Lamb, F.K., arXiv:1904.08907 [astro-ph.HE] (2019).

Misner, C.W., Thorne, K.S., and Wheeler, J.A., *Gravitation* (W. H. Freeman, San Francisco) (1973).

Montana, G., Tolos, L., Hanauske, M., and Rezzolla, L., *Phys. Rev. D* **99**, 103009, doi:10.1103/PhysRevD.99.103009, arXiv:1811.10929 [astro-ph.HE] (2019).

Most, E.R., Weih, L.R., Rezzolla, L., and Schaffner-Bielich, J., *Phys. Rev. Lett.* **120**(26), 261103, doi:10.1103/PhysRevLett.120.261103, arXiv:1803.00549 [gr-qc] (2018).

Na, X., Xu, R., Weber, F., and Negreiros, R., *Phys. Rev. D* **86**, 12, doi:10.1103/PhysRevD.86.123016 (2012).

Oppenheimer, J.R. and Volkoff, G.M., *Phys. Rev.* **55**, 374, doi:10.1103/PhysRev.55.374 (1939).

Orsaria, M., Rodrigues, H., Weber, F., and Contrera, G.A., *Phys. Rev. C* **89**(1), 015806, doi:10.1103/PhysRevC.89.015806, arXiv:1308.1657 [nucl-th] (2014).

Paschalidis, V., Yagi, K., Alvarez-Castillo, D., Blaschke, D.B., and Sedrakian, A., *Phys. Rev. D* **97**(8), 084038, doi:10.1103/PhysRevD.97.084038, arXiv:1712.00451 [astro-ph.HE] (2018).

Poghosyan, G.S., Grigorian, H., and Blaschke, D., *Astrophys. J.* **551**, L73, doi:10.1086/319851, arXiv:astro-ph/0101002 [astro-ph] (2001).

Raithel, C.A., Ozel, F., and Psaltis, D., *Astrophys. J.* **831**(1), 44, doi:10.3847/0004-637X/831/1/44, arXiv:1605.03591 [astro-ph.HE] (2016).

Ravenhall, D.G. and Pethick, C.J., *Astrophys. J.* **424**, 846, doi:10.1086/173935 (1994).

Read, J.S., Lackey, B.D., Owen, B.J., and Friedman, J.L., *Phys. Rev. D* **79**, 124032, doi:10.1103/PhysRevD.79.124032, arXiv:0812.2163 [astro-ph] (2009).

Rezzolla, L., Most, E.R., and Weih, L.R., *Astrophys. J.* **852**(2), L25, doi:10.3847/2041-8213/aaa401, arXiv:1711.00314 [astro-ph.HE] (2018).

Ropke, G., Blaschke, D., and Schulz, H., *Phys. Rev. D* **34**, 3499, doi:10.1103/PhysRevD.34.3499 (1986).

Ruiz, M., Shapiro, S.L., and Tsokaros, A., *Phys. Rev. D* **97**(2), 021501, doi:10.1103/PhysRevD.97.021501, arXiv:1711.00473 [astro-ph.HE] (2018).

Schaeffer, R., Zdunik, L., and Haensel, P., *Astron. Astrophys.* **126**, 121 (1983).

Sedrakyan, D.M. and Chubaryan, E.V., *Astrophys.* **4**, 227, doi:10.1007/BF01013134 (1968a).

Sedrakyan, D.M. and Chubaryan, E.V., *Astrophys.* **4**, 87, doi:10.1007/BF01020005 (1968b).

Seidov, Z.F., *Soviet Astronomy* **15**, 347 (1971).

Shibata, M., Fujibayashi, S., Hotokezaka, K., Kiuchi, K., Kyutoku, K., Sekiguchi, Y., and Tanaka, M., *Phys. Rev. D* **96**(12), 123012, doi:10.1103/PhysRevD. 96.123012, arXiv:1710.07579 [astro-ph.HE] (2017).

Shibata, M., Zhou, E., Kiuchi, K., and Fujibayashi, S., arXiv:1905.03656 [astro-ph.HE] (2019).

Spinella, W.M., Weber, F., Contrera, G.A., and Orsaria, M.G., *Eur. Phys. J. A* **52**(3), doi:10.1140/epja/i2016-16061-x (2016).

Stergioulas, N. and Friedman, J., *Astrophys. J.* **444**, 306, doi:10.1086/175605, arXiv:astro-ph/9411032 [astro-ph] (1995).

Tolman, R.C., *Phys. Rev.* **55**, 364, doi:10.1103/PhysRev.55.364 (1939).

Weber, F. and Glendenning, N.K., *Astrophys. J.* **390**(2), 541, doi:10.1086/171304 (1992).

Weber, F., *Pulsars as Astrophysical Laboratories for Nuclear and Particle Physics* (Taylor & Francis, 1999).

Yagi, K. and Yunes, N., *Phys. Rev. D* **88**(2), 023009, doi:10.1103/PhysRevD.88. 023009, arXiv:1303.1528 [gr-qc] (2013).

Yakovlev, D.G., Kaminker, A.D., Gnedin, O.Y., and Haensel, P., *Phys. Rept.* **354**, 1, doi:10.1016/S0370-1573(00)00131-9, arXiv:astro-ph/0012122 [astro-ph] (2001).

Yasutake, N., Lastowiecki, R., Benic, S., Blaschke, D., Maruyama, T., and Tatsumi, T., *Phys. Rev. C* **89**, 065803, doi:10.1103/PhysRevC.89.065803, arXiv:1403.7492 [astro-ph.HE] (2014).

Zdunik, J.L., Bejger, M., Haensel, P., and Gourgoulhon, E., *Astron. Astrophys.* **450**, 747, doi:10.1051/0004-6361:20054260, arXiv:astro-ph/0509806 [astro-ph] (2006).

Zdunik, J.L. and Haensel, P., *Astron. Astrophys.* **551**, A61, doi:10.1051/ 0004-6361/201220697, arXiv:1211.1231 [astro-ph.SR] (2013).

Index

www.ingramcontent.com/pod-product-compliance
Lightning Source LLC
Chambersburg PA
CBHW050544190326
41458CB00007B/1916